# ENERGY TECHNOLOGY SERIES

# GASOHOL
## for
# ENERGY PRODUCTION

BY

## NICHOLAS P. CHEREMISINOFF

ANN ARBOR SCIENCE
PUBLISHERS INC
P.O. BOX 1425 • ANN ARBOR, MICH. 48106

# GASOHOL
## for
# ENERGY PRODUCTION

## ENERGY TECHNOLOGY SERIES

**GASOHOL FOR ENERGY PRODUCTION
WOOD FOR ENERGY PRODUCTION
BIOGAS PRODUCTION AND UTILIZATION
FUNDAMENTALS OF WIND ENERGY
PRINCIPLES AND APPLICATIONS OF
  SOLAR ENERGY**

Additional volumes will cover coal, hydrogen,
hydroelectric and geothermal energy generation,
ocean thermal energy conversion, and insulation.

# FOREWORD

During the past century and a half, and particularly in the last two to three decades, there has been a tremendous increase in uses and demands for energy for transportation, industry and personal needs. This energy has been supplied by our stocks of coal, oil and natural gas, all of which have become increasingly costly, scarce and environmentally questionable.

In this last quarter of the twentieth century our society is now looking to new or alternate energy technologies, such as solar, geothermal, nuclear, wind and biomass. These alternate energy sources may well provide the ingenious solutions we are seeking. If we look not too far back in history, during the 1940s, buses in China were powered by alcohols; Germany, Britain and France powered cars during World War II by liquid fuels made from coal and coal gas.

The art or science of producing alcohols from grains and fruits dates back to ancient times. Egyptians and Mesopotamians recorded methods to brew alcohols for consumption as early as 2500 B.C. Alcohols have also been used as fuels on a limited and intermittent basis. Today, with petroleum shortages and subsequent rising costs, interest is returning.

There are a variety of sources from which alcohol fuels can be derived, and biomass represents a renewable resource capable of providing safe nonpolluting fuels. Biomass materials can alternately be the source of food, materials and/or chemicals. Even when biomass is useful solely as a feedstock for energy generation, the principles of its production can be applied to produce food, animal feed, textiles, paper, chemicals, etc.

Mankind is experiencing the end of an era—an era of cheap, abundant energy. At our current rate of energy consumption, within a few short decades remaining petroleum and natural gas supplies may become exhausted.

This book was written as an overview of the present state-of-the-art and the potentials and uses of biomass as a source of alcohols and chemical feedstocks. A generous reference list and bibliography is provided for those who may wish to probe deeper into this subject.

Nicholas P. Cheremisinoff

 **Nicholas P. Cheremisinoff** received his BS, MS and PhD degrees in chemical engineering from Clarkson College of Technology, where he was instructor during 1976-1977. He is currently at Exxon Research and Engineering Co. in Florham Park, New Jersey, and is Adjunct Professor at New Jersey Institute of Technology in the Civil & Environmental and Chemical Engineering departments. Prior to Exxon he was Research Scientist at Union Camp R&D Division in Princeton, New Jersey. Dr. Cheremisinoff has contributed to industrial press and has authored sections in several engineering books. He is a member of a number of professional societies including Tau Beta Pi, Sigma Xi and AIChE.

# TABLE OF CONTENTS

## BIOMASS—A SOURCE OF ENERGY

### INTRODUCTION

During the past decade, man has become increasingly aware of the limitations of energy supplies. With further increases in energy consumption expected, we are faced with the question of how long there will be sufficient reservoirs of conventional fuels for our industrialized societies? The increasing scarcity of fossil fuels and the attendant increase in the cost of petroleum products make it imperative to find alternate energy sources during the next decade.

Alternate energy sources must meet certain criteria to be competitive with conventional fuels. Some of the special requirements these energy supplies will have to meet are as follows:

1. Fuels must be capable of being stored over extended time periods,
2. Storage, transportation and distribution of fuels used should be economical,
3. Handling of alternate fuels should not involve additional hazards such as fire, explosion, etc., in comparison to conventional fuels,
4. Alternate fuels should not impose major engineering changes to processes and/or systems using them.

Recently, considerable attention has been given to alcohol fuel production from biomass. A specific application that this book addresses is the possibility of blending biomass-derived alcohol with gasoline to produce a transportation fuel. It is technologically possible to develop a liquid transportation fuel system based on alcohol-gasoline blends.

The use of alcohol-based fuels is not a new concept. In fact, alcohols have been utilized extensively throughout the world as petroleum substitutes during periods of shortage. The relative abundance and low cost of petroleum since the end of World War II reduced the importance of alcohol fuels.

Today's petroleum shortages have created renewed interest in alcohol fuel. This has stimulated new research into applications and bioconversion techniques, as well as reassessment of the environmental advantages/disadvantages of using alcohol fuels.

## HISTORICAL DEVELOPMENTS

It was only 300 years ago that man first began to efficiently harness energy by converting solid fossil fuels to mechanical energy with primitive steam engines. America followed England's and Germany's thrusts of mechanization of industry and transportation through the use of coal. Coal successfully supplied man's energy needs for more than 200 years.

Petroleum has been recognized as an energy supply since ancient times; however, 1859 marked the beginning of a new era in the use of energy with Drake's Discovery Well in Pennsylvania. This represented the first significant source of oil in America.

Low-pressure gas wells were also discovered during this time period. Natural gas was first utilized in homes and found limited use in various industries. Traditionally, gas has been among man's most easily controlled and most readily transported fuels, with large pipelines stretching over thousands of miles. It was also the last fossil fuel to experience wide use in this country.

Until recently, the abundance of oil established prices for petroleum below its true energy value to man. In this century, man has consumed more energy than in his entire history on earth, with the bulk of it coming from fossil fuels.

Within the next few decades our natural gas and petroleum supplies may be reduced to such an extent that if short- as well as long-term solutions are not found, many may have to adjust to radical changes in lifestyle and technologies.

A vast source of energy relatively untapped by man thus far is biomass. Lignocellulose and various other carbohydrates were transformed into fossil fuels millions of years ago, thus forming the basis of our present-day energy resources. Their growth in vegetation is considered as the most efficient converter of solar energy. Biomass, which takes many forms, has been recognized and used as an energy source since ancient times. In this country, for example, wood was our primary energy source about 100 years ago, supplying roughly 75% of the U.S. energy consumption. Since that time, U.S. energy consumption has increased from 4 Q (1 Q = 1 quad = $10^{15}$ Btu) to 75 Q, with total wood fuel usage only contributing about 1.1 Q today.[1]

Both biological and thermochemical conversion processes can be employed for converting various biomass feedstocks to alcohols having fuel values. Considerable attention has been directed at the near-term possibility of blending biomass-derived alcohol with gasoline. The earliest recorded use of alcohol as a motor fuel is 1890. Ethanol and methanol, both monohydric alcohols, found extensive use in Europe during the pre-World War II era.[2] In this country, alcohol fuels have been limited to special fuel mixtures for racing car engines. Mixtures of water and alcohol have been employed in injection into high-compression aircraft engines.

## SOURCES OF BIOMASS FEEDSTOCKS

Cellulose comprises roughly 30-50% of all vegetation, and although large amounts are used in the form of paper, lumber, textiles and feed, large quantities are left unused. This unused cellulose is left to undergo natural decay or is discarded in the form of domestic, industrial or agricultural wastes. These biological wastes represent roughly $1.52 \times 10^{14}$ $m^3$ of natural gas and in terms of economics, $500 billion to $1 trillion annually.[3,4] Table 1.1 gives a rough breakdown of biomass wastes.

Table 1.1 Estimates of Solid Wastes Suitable for Biomass Feedstock in the U.S.[3,4]

| Biomass Source | Estimated Wastes ($10^6$ tonne/yr)[a] | | Total Waste | Water Content |
|---|---|---|---|---|
| | Dry Basis | Wet Basis | % | % |
| Urban Refuse | 117.9 | 235.9 | 14.7 | 66.7 |
| Crop Residues | 353.8 | 499.0 | 44.3 | 58.5 |
| Manures | 181.4 | 1361-1814 | 22.7 | 88-91 |
| Industrial Wastes | 39.9 | 99.8 | 5.0 | 71.4 |
| Logging Wastes | 49.9 | 72.6 | 6.2 | 59.3 |
| Sewage | 10.9 | – | 1.4 | – |
| Miscellaneous | 45.4 | – | 5.7 | – |
| Total Waste | 799.2 | 2268-2722 | 100 | 74-78 |

[a]1 tonne = metric ton = 1.1023 x short ton.

As shown in Table 1.1, urban refuse, crop residues and manures constitute the major sources of organic wastes. Of all the materials listed, only urban refuse is collected routinely. Crop and logging residues are collected seasonally but not on an organized basis. More than 80% of the total waste disposal costs can be attributed to collection; consequently, for many seasonal wastes it is not economical to collect and dispose of them. This is also the reason why urban wastes are routinely disposed of near their collection sites.

Crop residues represent more than 44% of the total estimated residue in the U.S. More than 95% of the field crops planted are for food grains.[5] The majority of the residuals (*e.g.,* corn stalks, wheat straw, etc.), are left unused and to decompose once the crop has been harvested. More than 350 tonnes of dry organic residue are generated annually (Table 1.1). Clausen *et al.*[6] notes that if all these residues could be collected for energy production, roughly 6% of the total U.S. energy consumption could be supplied. Although this is significant, its total impact on the energy picture is small. The amount of waste generated depends on a number of factors, including soil classification, geographic location, average climatic conditions, seasonal variations and type of crop. Wastes for most crops are presently collected on the farm and plowed back into the soil or, in some cases, used for fodder.

The amount of waste generated from urban areas is generally a function of population size. Larger cities do tend, however, to generate more commercial refuse per capita than smaller cities. Urban wastes vary greatly in composition. In general, roughly 50% of the bulk weight tends to be inorganic matter and water. The remaining portion represents dry organic material having a potential energy value.

Among industrial wastes, the lumber and food processing industries are the largest contributors of organic matter suitable for energy production. The chemical and manufacturing industries on the whole generate low amounts of organic matter. Food processing wastes vary greatly in both composition and moisture content. Roughly 30% or more of these residues are used in by-products. The remaining wastes, which are disposed of, generally have low heating values in the range of 11,600-18,600 abs J/kg.

Wastes from logging operations include deadwood, culls, brush and branches. Normal practice is to leave the debris where it falls, chip it to accelerate decomposition and/or burn it at the site. The environmental disadvantages associated with current disposal methods include fire, air and ground-water pollution.

The amount of waste generated from the logging industry is significant. One study[7] has indicated that approximately 1 kg of waste debris is generated for each board of harvested logs. Additional wood wastes are generated from sawmills and planing mills. Mill residues are already being used as an energy feedstock in the paper industry. The majority of this residue is usually burned at the mill site (normally in a power boiler), while a portion of the remainder is used in by-products, such as pressed logs, or as part of the fill back to digesters for pulp production.

The heating value of mill and logging residues varies with wood species, as well as the relative contents of wood and bark, and moisture content. Inman[8] reports that roughly 95 million tonnes (dry weight basis) of mill residue were generated in 1970 alone. Figure 1.1 illustrates the relative contributions of these wastes from various industries. Roughly 56% of all of these residues were used for nonenergy products.

The term "forest residues" includes logging residues, intermediate cuttings, understory and annual mortality, *i.e.,* trees killed by natural causes. Intermediate cuttings refers to the removal of small or inferior trees from a stand for the purposes of stand improvement. Understory removal involves the removal of shade-tolerant shrubs and trees that grow underneath the canopy of an older commercial forest. Both these sources represent significant supplies of biomass suitable for energy production.

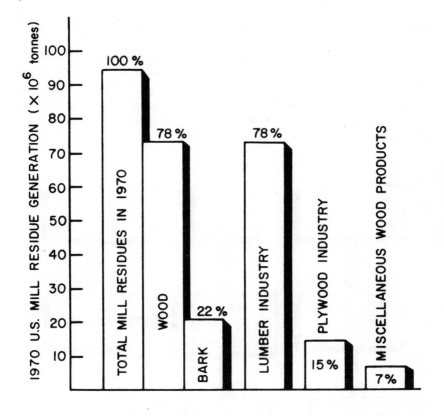

Figure 1.1 The relative contributions to total mill residues produced in 1970 from various segments of the logging industry.[8,9]

Logging residues, leftovers of commercial logging operations, are considered to be the aboveground portions of trees not removed in these operations, *e.g.,* branches, foliage or any portion of the tree that is defective or not suitable for conventional wood processing. Forest residues presently do not have any value to forest landowners.

Animal wastes present a serious disposal problem and one which is becoming more acute due to the large number of animals being bred on concentrated feedlots. Standard disposal methods include overland spreading, burying, incineration or collecting in piles. Because none of these methods is environmentally suitable, energy production alternatives are promising.

Each of the waste materials discussed briefly has potential energy benefits for man. Forest residues and agricultural wastes constitute significant biomass resources that are suitable for alcohol production. The technology for alcohol production is well defined and, in this country, has been well developed by the chemical industry.

Alcohol as a fuel offers promising alternatives to fossil-based liquid fuels. Various biomass fuel production routes and biomass feedstocks are illustrated in Figure 1.2. Methanol and ethanol are considered to be the most promising alcohols, suitable for extensive use as transportation fuels. However, as will be discussed in subsequent chapters, a variety of other alcohols and energy-related applications can be derived from biomass.

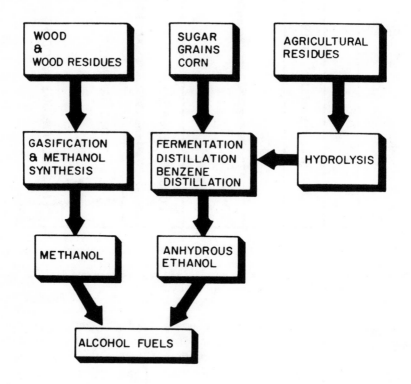

Figure 1.2 Illustrates general approaches to the production of methanol and ethanol.

## 2

# THE CHEMISTRY OF ALCOHOLS

## GENERAL DESCRIPTION

Alcohols are basically any class of chemical compounds consisting of an alkyl group, derived from a hydrocarbon by the replacement of a hydrogen atom by an hydroxyl radical. The general formula used for denoting an alcohol is ROH, where R signifies an alkyl group and OH the hydroxyl radical. Simple alcohols are named after the alkyl, cycloakyl or alkenyl groups to which the hydroxyl groups are bonded. Figure 2.1 gives several examples of alcohols.

Alcohols are also called carbinols. For example, methyl alcohol can be simply called carbinol. There are a number of alcohols whose names are convenient derivatives of carbinol. The hydrocarbon groups are considered substitutes for one or more of the hydrogen atoms of carbinol. This approach is based on classifying alcohols as primary, secondary or tertiary, depending on whether one, two or three of the hydrogens of carbinol are replaced by hydrocarbon groups. Figure 2.2 illustrates the system of classifications.

The name glycol refers to an aliphatic or alicyclic compound containing two hydroxyl groups. Aliphatic and alicyclic compounds with three hydroxyl groups are referred to as trihydric alcohols. If they contain two or more, they are called polyhydric alcohols. Other common names used for these compounds are diols, triols or polyols.

In the International Union of Chemistry system (IUC), names of compounds are based on the premise that each organic compound has a basic skeletal structure, called the parent, to which functional groups and branches have been added. The nomenclature of an alcohol based on IUC uses the longest carbon chain to which the hydroxyl is attached as the parent compound, and the chain is numbered from the end closest to the hydroxyl group. The presence of two or more hydroxyl groups is denoted by the suffixes -diol, -triol, etc.

Greek letters are used to indicate the position of substituents on saturated hydrocarbon chains. For example, the $\alpha$-carbon refers to the carbon atom bound to the hydroxyl group.

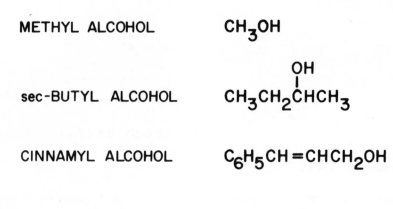

METHYL ALCOHOL          $CH_3OH$

sec-BUTYL ALCOHOL       $CH_3CH_2\overset{\overset{\displaystyle OH}{|}}{C}HCH_3$

CINNAMYL ALCOHOL        $C_6H_5CH = CHCH_2OH$

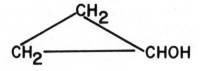

CYCLOPROPYL ALCOHOL

$C_6H_5CH_2OH$          BENZYL ALCOHOL

Figure 2.1 Typical examples of simple alcohols.

## BIOMASS-BASED ALCOHOLS AND THEIR PROPERTIES

Methanol and ethanol are currently considered the most potentially viable alcohols for extensive uses as transportation fuels, although many alcohols have considerable commercial importance. Methanol is a clear, toxic alcohol that is commonly referred to as wood alcohol. At one time, methanol was produced entirely from the destructive distillation of wood. Today, it is produced in large quantities from carbon monoxide and hydrogen and finds greatest use as a solvent.

Methanol production from biomass is carried out in a two-step process (described later in detail). The first part of the process is a high-temperature gasification process, which results in a synthetic gas. The second step involves the liquefaction of the synthetic gas and final catalytic conversion to methanol.

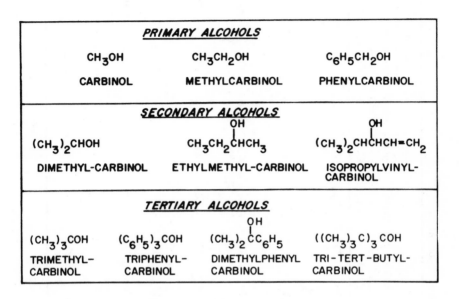

Figure 2.2  Classification of alcohols.

Ethanol has enjoyed a long history as a beverage.  As a beverage it is derived entirely from the fermentation of grains and various sugar or starch feedstocks.  Industrial-grade ethanol is primarily manufactured from ethylene (a gas derived from petroleum).

Another alcohol produced by fermentation is *n*-butyl alcohol (1-butanol). Like ethanol, it is widely used as a solvent.  Ethanol, 2-propanol and diethyl ether are manufactured from ethylene feedstock.

In general, alcohols are like water in that they are amphoteric and are neither strong bases nor strong acids. The order of acidity for alcohols is normally primary > secondary > tertiary.  As an example, *t*-butyl alcohol is much less acidic than ethanol.

Alcohols are considered bases, comparable in strength to water. They can be converted into more or less stable salts with strong acids.  Basic physiochemical properties of different alcohols are summarized in Table 2.1. As seen from Table 2.1, melting points generally tend to increase with molecular weight.  Exceptions to this are methyl and ethyl alcohols, as these melt at a slightly higher temperature than propyl alcohol.

Alcohols are colorless compounds. Their solubility in water decreases sharply with increasing molecular weight. Boiling points tend to increase with molecular weight. A more detailed description of the specific alcohols follows.

Table 2.1  Physiochemical Properties of Various Alcohols[2,10]

| Chemical Compound | Formula | Description | Molecular Weight | Refractive Index T(°C) | Refractive Index Index | Melting Point (°C) | Boiling Point (°C) | Flash Point (°C) | Specific Gravity Reference to Air (°C/°C) | Specific Gravity Ratio |
|---|---|---|---|---|---|---|---|---|---|---|
| Methyl Alcohol (Methanol) | $CH_3OH$ | Colorless liquid | 32.042 | 20 | 1.329 | -97.8 | 64.5 | 12.2 | 20/4 | 0.792 |
| Ethyl Alcohol (Ethanol) | $C_2H_5OH$ | Colorless liquid | 46.070 | 15 | 1.363 | -116. | 78.4 | 13.9 | 20/20 | 0.790 |
| n-Propyl Alcohol (Propanol-1) | $CH_3CH_2CH_2OH$ | Colorless liquid | 60.097 | 20 | 1.385 | -126.1 | 97.3 | 15.0 | 20/4 | 0.840 |
| Isopropyl Alcohol (2-propanol) | $CH_3CHOHCH_3$ | Colorless liquid | 60.097 | 20 | 1.378 | -85.8 | 82.4 | 11.7 | 20/4 | 0.787 |
| n-Butyl Alcohol | $C_2H_5CH_2CH_2OH$ | Colorless liquid | 74.124 | 25 | 1.397 | -79.9 | 117.7 | 28.9 | 20/20 | 0.811 |
| sec-Butyl Alcohol | $C_2H_5CHOHCH_3$ | Colorless liquid | 74.124 | 20 | 1.397 | -114.7 | 98.8 | 21.1 | 20/20 | 0.808 |
| Isobutyl Alcohol | $(CH_3)_2CHCH_2OH$ | Colorless liquid | 74.124 | 20 | 1.396 | -108. | 108.3 | 27.8 | 20/20 | 0.803 |
| tert-Butyl Alcohol | $(CH_3)_3COH$ | Colorless liquid | 74.124 | – | 1.388 | 25.6 | 82.4 | 11.1 | 25/25 | 0.783 |

## METHANOL

Methanol is rarely found in its free state in nature; however, derivatives of methanol are fairly common. There are a variety of plant oils that contain methyl esters. Examples include oil of jasmine, which contains methyl anthranilate ($C_6H_4$ ($NH_2$) $\cdot$ $COOCH_3$), and oil of wintergreen, which contains methyl salicylate ($C_6H_4$ (OH) $\cdot$ $COOCH_3$).

At one time, methanol was manufactured entirely from the destructive distillation of wood. Today the most commonly practiced method involves reducing carbon monoxide to methyl alcohol by hydrogen in the presence of mixtures of metal oxides, which serve as catalysts. Both processes are described later. A by-product formed from the latter process includes isobutyl alcohol.

The hydrolysis of an ester produces the best results when chemically preparing pure methanol. In its pure state, this alcohol is a colorless liquid with an odor similar to ethanol. Methanol is an intoxicant and a strong poison.

Its chief industrial use is as a solvent for lacquers and perfumes. It is also used in the production of mono- and dimethylaniline; in the manufacture of methyl chloride and dimethyl sulfate; for the production of formaldehyde; and for denaturing spirits.

Methanol has been proposed for direct use as a substitute fuel in automobile engines. If this route is taken, however, carburetion systems currently in use would require major modifications. Engines designed to operate on gasoline alone will not function on pure alcoholic fuel without significant changes.

Methanol can be mixed with gasoline in various proportions. There are problems that arise with the use of methanol-gasoline blends for automobile fuels.

## ETHANOL

Ethanol, or ethyl alcohol, is formed in nature by the fermentation of carbohydrates. It exists in trace quantities in plant life, animal tissues and blood.

There are a variety of methods that can be used to synthesize ethanol; however, only a few have been adapted as industrial processes. Good results are obtained from the synthesis of acetylene. In this approach, acetaldehyde is formed as an intermediate, which is catalytically reduced to ethanol by hydrogen in the presence of nickel. The most economical process for ethanol production is, however, alcoholic fermentation.

Ethanol has found wide use both in industry and in the laboratory, as a solvent and extracting agent. It is useful as a fuel, in the preparation of pharmaceuticals and perfumes, for the production of acetic acid, various lacquers, varnishes and dyes. The largest market for ethanol is as a beverage, partly in the form of liquors (manufactured artificially from pure alcohol, water, sugar and essences) and partly as alcoholic drinks. The latter is made from raw materials, consisting primarily of starches or sugars. Strong alcoholic beverages are obtained by performing distillation after fermentation. Table 2.2 gives specific examples of various raw feedstocks that go into the production of different alcoholic beverages.

Table 2.2  Natural Feedstocks used in Alcoholic Beverage Production

| Raw Material | Basic Constituent | Alcoholic Product | |
| --- | --- | --- | --- |
| | | Fermentation Alone | With Distillation |
| Grapes | Sugar | Wine | Brandy |
| Fruit | Sugar | Cider | |
| Molasses | Sugar | | Rum, arrack |
| Cherries | Sugar | | Cherry brandy |
| Plums | Sugar | Plum wine | Slivovic |
| Wine and Cider | | | |
| Residues | Sugar | | Weak spirits |
| Barley | Starch | Beer | Corn brandy |
| Wheat | Starch | Beer | Corn brandy |
| Rye | Starch | Beer | Whiskey, corn brandy |
| Potatoes | Starch | | Brandy |
| Rice | Starch | Saki | Arrack |

## BUTYL ALCOHOLS

There are basically four possible structurally isomeric butyl alcohols. Their chemical formulas and names are given in Figure 2.3.

Normal butyl alcohol is produced commercially from starch, and starchy materials by fermentation with *B. acetobutylicus,* which consists of about 60% butyl alcohol, 10% ethanol and 30% acetone.[11] This alcohol has found widest usage as a solvent for nitrocellulose lacquers.

Secondary butyl alcohol is produced from the reduction of methyl ethyl ketone. It can also be prepared from normal butylenes.

Primary isobutyl alcohol is obtained from fractional distillation of fuel oil. The free oil can also be found in distillate of wood. It is often produced commercially from water gas ($CO$ & $H_2$) via cobalt salts catalysts. Isobutyl alcohol is used in perfumery, as a solvent for lacquers, and in the synthesis of pharmaceuticals.

Tertiary butyl alcohol is made from isobutylene and from acetone and methyl-magnesium salts (Grignard).

| STRUCTURAL FORMULA | NAME |
| --- | --- |
| $CH_3CH_2CH_2CH_2OH$ | BUTANOL-1 (PRIMARY) NORMAL BUTYL ALCOHOL |
| $CH_3CH_2CHOHCH_3$ | BUTANOL-2 ; SECONDARY BUTYL ALCOHOL |
| $CH_3CHCH_2OH$<br>$\quad\ \ \vert$<br>$\quad\ \ CH_3$ | METHYL-PROPANOL-1 (PRIMARY) ISOBUTYL ALCOHOL |
| $CH_3C(OH)CH_3$<br>$\quad\ \vert$<br>$\quad\ CH_3$ | METHYL-PROPANOL-2 ; TERTIARY BUTYL ALCOHOL |

Figure 2.3  The butyl alcohols.

## PROPYL ALCOHOLS

There are two monohydric propyl alcohols of importance, namely *n*-propyl alcohol and isopropyl alcohol. The former is manufactured from ethylene gas via oxo technology, or by an oxidation process using propane. *N*-propyl alcohol can also be obtained in small quantities as a by-product from the distillation of fermented products. *N*-propyl alcohol is employed primarily as a solvent and as a chemical intermediate.

Isopropyl alcohol is generated from the hydration of propylene. As in ethanol production, final distillation of isopropyl alcohol produces a binary mixture of water and the alcohol (roughly 91% isopropyl on a volume basis). At this purity, the alcohol can either be sold or further dehydrated through the use of an azeotropic constituent, *e.g.,* benzene or isopropyl ether. Isopropyl alcohol is used in coatings, as a solvent, as an intermediate in drugs and cosmetics manufacturing, and in the production of acetone.

## 3

## METHANOL SYNTHESIS

**EARLY DEVELOPMENTS**

Charcoal is produced from the thermal decomposition of wood at temperatures in the range of 160-430°C, in the absence of air (called pyrolysis). There are several by-products generated from charcoal production; these include wood tar, noncondensible gases and pyroligneous acid (a watery distillate). Pyroligneous acid can be distilled or subject to extraction to recover methanol, as well as acetic acid and acetone. For hardwood species, a very low yield is obtained, producing only 1-2% methanol (or roughly 2.3 l/tonne).[12] For softwood species, considerably less than this yield is achieved.[13]

As early as 1905, a French chemist, Paul Sabatier, proposed the synthesis of methanol via hydrogenating carbon monoxide. In 1921, the Haber process was perfected, whereby methanol was synthesized from carbon monoxide and hydrogen. M. Patart patented the process for a temperature range of 300-600°C, with a pressure of $1.5 \times 10^7$ to $2.0 \times 10^7$ $N/m^2$, *i.e.*, 150-200 atm.[12,14]

The first commercial synthetic methanol plant was built in Leunawerke, Germany in 1923. This had a dramatic impact on the wood chemicals industry. In 1924, synthetic methanol began being exported to the United States at a fraction of the cost that existing wood-derived methanol plants could produce.

Synthetic processes were shortly adopted in the U.S. Commercial Solvents Corporation began manufacturing synthetic methanol from hydrogen and carbon monoxide during the fermentation of corn in 1926. E. I. du Pont de Nemours & Co. developed a similar process using carbon monoxide and hydrogen derived from ammonia synthesis.[15]

## METHANOL PRODUCTION FROM SYNTHESIS GAS

The primary source of methanol today is natural gas. Basically, any material that can be thermally decomposed into hydrogen and carbon monoxide or carbon dioxide, *i.e.*, the synthesis gas, is considered a potential source of methanol.

The majority of methanol produced is made by passing the synthesis gas over a catalyst at elevated temperatures and under pressure. Temperatures normally range up to $400°C$, with pressure between $5.07 \times 10^6$ to $3.55 \times 10^7$ $N/m^2$.

Synthesis gas (SNG), or more recently called Syngas, is often produced from the gasification of fossil fuels. The gas must be purged of sulfur to protect against any interaction with the synthesis catalyst. It is then reacted with steam to promote a water-gas shift reaction. (This shifts the ratio of gases toward a stoichiometric ratio of 2:1 to 3:1.)

There are two primary reactions that can be used in the production of methanol. The synthesis gas feedstock approach is defined by Equations 3.1 and 3.2:

$$CO_2 + 3H_2 \longrightarrow CH_3OH + H_2O \qquad (3.1)$$

$$CO + 2H_2 \longrightarrow CH_3OH \qquad (3.2)$$

The synthesis gas can be made from a variety of carbonaceous sources, although considerable interest has been given to coal gasification. The following equations describe the generation of the synthesis gas:

$$C + H_2O \longrightarrow CO + H_2 \qquad (3.3)$$

$$C + 2H_2O \longrightarrow CO_2 + 2H_2 \qquad (3.4)$$

Figure 3.1 illustrates the basic process for methanol production from SNG. The synthetic methanol method is essentially a low-conversion process. To obtain a substantial yield, a relatively large recycle stream is required. The converter shown in Figure 3.1 has two feed stations. Feed stream (B) is basically a bypass stream, which provides for flow control through the reactor. This allows temperature adjustments in the converter bed.

## METHANOL PRODUCTION FROM HYDROCARBON OXIDATION

Another successful method for methanol production is hydrocarbon oxidation. In Equation 3.5, methane is oxidized to methanol:

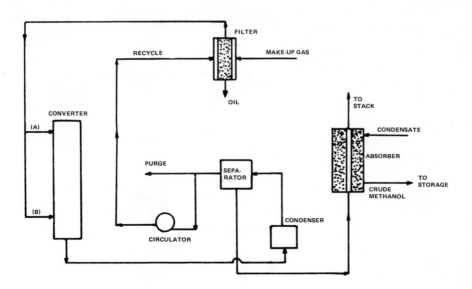

Figure 3.1  A medium-pressure methanol synthesis plant.

$$2CH_4 + O_2 \longrightarrow 2CH_3OH \qquad (3.5)$$

A variety of hydrocarbons can be oxidized to produce methanol; however, considerably lower yields are obtained from larger hydrocarbons. In general, the large hydrocarbon molecules tend to be only partially oxidized, resulting in smaller methanol yields.

Methane and other light hydrocarbons can also be used to generate synthesis gas. This is described by the following group of reactions:

$$CH_4 + H_2O \longrightarrow CO + 3H_2 \qquad (3.6)$$

$$CH_4 + 2H_2O \longrightarrow CO_2 + 4H_2 \qquad (3.7)$$

$$CO + H_2O \longrightarrow CO_2 + H_2 \qquad (3.8)$$

Equation 3.8 is referred to as the "shift reaction."

Figure 3.2 illustrates the methane oxidation process. The use of catalysts in the oxidation furnace is a somewhat controversial issue. Several catalysts that have proved successful are iron salts, nickel, copper, palladium, various metal oxides, mixtures of these oxides, aluminum sulfate and alkyl ethers. In principle, at least, any hydrocarbon can be partially oxidized to produce

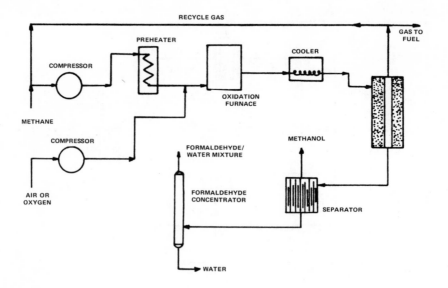

Figure 3.2  The process for methanol production via partial oxidation of methane gas.

methanol; however, the yield will decrease as the carbon chain of the feed-stock increases.  It is important, therefore, that a highly selective catalyst be used for large carbon number feedstocks.

Oxidation of methane by sulfur trioxide in liquid sulfuric acid is one approach that has been proposed.[15]   Another process that has not been used commercially is the saponification of methyl chloride by caustic soda. Methyl chloride can be produced by chlorinating methane with chlorine or hydrogen chloride and oxygen.

## THERMODYNAMIC AND CHEMICAL CONSIDERATIONS FOR METHANOL SYNTHESIS

Aharoni and Starer[16] attempted to interpret the reactions of hydrogen and carbon monoxide on zinc-chromium catalyst surfaces.  Their study indicated that surface hydrogenation of carbon monoxide takes place in two stages.  The first stage involves the formation of an intermediated compound, which takes place over a relatively wide range of temperatures and absorbate compositions.  Possible intermediate structures envisioned are illustrated in Figure 3.3.  Methanol can be formed by the breaking of the compound surface bond between either intermediate and promoting hydrogenation.  If the carbon-to-oxygen bond is broken and hydrogenation occurs, methane is formed. Water is also generated during the latter reaction.

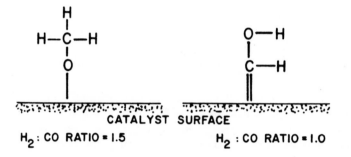

Figure 3.3 Two possible structures of the intermediate compound as envisioned by Aharoni and Starer.[16]

The second stage involves the formation of methanol or methane and a desorption step. Desorption conditions establish the selectivity of the reaction. It should be noted that methane could be desorbed irreversibly at temperatures that do not favor methanol desorption.

The use of catalysts greatly complicates the kinetics of the reactions. Many catalysts have been found to have high initial activities and suitable selectivity; however, many suffer from short activity. Most catalysts are highly sensitive to poisoning from sulfur; hence, there is a need for adequate purging of the synthesis gas. Zinc and chromium oxide catalysts tend to have lower activities in comparison to copper-based catalysts; however, they tend to have longer life cycles. Relatively high temperatures and pressures are needed to obtain sufficiently fast reaction rates with these catalysts. For example, Hagen[12] notes that a 72% zinc oxide, 28% chromium oxide catalyst has an optimum activity at about 375°C. Other metal oxides are often added as filler material to promote activity.

Low-pressure methanol processes produce yields typically in the range of 2.5% in the exit gas. By contrast, a high-pressure process might generate as high as 5.5% yield. Although a low-pressure system involves considerably higher recirculation of unreacted gases to maintain an economical yield, a significant energy savings is realized because of the large reduction in compression requirements of the synthetic gases. Specifically, with a low-pressure system, a 40% reduction in pumping energy is achieved over a high-pressure process (combined compression and circulation is typically 270 kWh/ton for a low-pressure process, whereas a large-pressure process is 470 kWh/ton).[17]

Methanol production efficiency is treated on a first law basis. That is, the enthalpies for the products are divided by the energy in the feedstocks and the energy inputs (Equation 3.9):

$$\epsilon = \Delta H^\circ_{prod}/(\Delta H^\circ_f + \Delta H^\circ_a) \qquad (3.9)$$

where $\epsilon$ is efficiency, $\Delta H^\circ_{prod}$ is the enthalpy for the products, and $\Delta H^\circ_f$ and $\Delta H^\circ_a$ are the enthalpies for the feedstocks and energy inputs, respectively. Methanol production efficiencies for large-scale gas plants have been reported in the range of 50-60%.[18,19] Methanol-producing plants based on municipal waste or wood are expected to have efficiencies in the range of 30-37%.

The definition of efficiency given in Equation 3.9 is based on first law thermodynamics. Efficiencies based on Equation 3.9 are misleading because there are limitations placed on thermodynamic conversion processes by the Carnot Cycle. Some investigators are therefore defining efficiencies in terms of the second law, using potential energy rather than the total energy of the first law for calculation purposes. Efficiency is then defined as the ratio of actual available or potential energy transferred to the products to the maximum theoretical amount that could be transferred. Using the Gibbs free energy of combustion, then:

$$\epsilon = \Delta G^\circ_c (products)/\Delta G^\circ_c (fuel) \qquad (3.10)$$

And the Gibbs free energy of combustion is defined as:

$$\Delta G^\circ_c = \Delta H^\circ_c - T\Delta S^\circ_c \qquad (3.11)$$

Equation 3.11 give values for the Gibbs free energy of formation, and combustion as derived from the literature for the chemical constituents encountered. In the equation, $\Delta H^\circ_c$ is enthalpy, T is temperature, and $\Delta S^\circ_c$ is entropy. Equations 3.10 and 3.11 result in calculated values for efficiency 3-4% higher than the definition based on first law principles.

Naturally, production efficiencies are highly dependent on the thermodynamic equilibria existing for a set of operating conditions. The equilibrium relationship existing between carbon monoxide, hydrogen and methanol is a function of temperature and the partial pressure of the gases. As an example, consider the reaction described by Equation 3.2. The equilibrium constant for this reaction is given as follows:

$$k_f = \frac{f_{CH_3OH}}{(f_{co})(f_{H_2})^2} = \exp(-\Delta G/RT) \qquad (3.12)$$

Parameter f is the fugasity of the constituent gas, and G is the Gibbs free energy. Woodward[15] gives the general form of the free energy expression (assuming a cubic heat capacity equation) as follows:

$$\Delta G^{\circ} = -74,622 + 67.28\, T\, \ell n(T) - 0.04682\, T^2 + 4.259 \times 10^{-6}\, T^3$$
$$+ 0.339 \times 10^{-9} T^4 - 202 T \qquad (3.13)$$

where T is temperature ($^{\circ}$K), and $\Delta G^{\circ}$ is in units of joules per gram-mole (J/g-mol).

The kinetics of the reaction, i.e., the rate at which conversion to the equilibrium concentration is approached, is complicated, especially with the use of catalysts. The overall reaction combines the rates of reaction of the synthesis gas as well as the degree of decomposition of methanol product, i.e., the overall rate of product reaction is a combination of the rates of reaction of the synthesis gas minus the rate of methanol decomposition. The reaction rate is, in reality, a complicated function of the catalyst composition and the composition and condition of the absorbed species. Additional complications in describing adequate rate relationships stem from the nonideality of the catalyst, variations in reactor temperature, and the rate of adsorption onto the catalyst surface.

The general expression describing the reaction rate for methanol synthesis was developed by Natta et al.[21]:

$$\theta = \frac{f_{co} f_{H_2}{}^2 - f_{CH_3OH}/k_p{}^{\circ}}{(\alpha + \beta f_{co} + \psi f_{H_2} + \phi f_{CH_3OH})} \qquad (3.14)$$

In Equation 3.14, $\theta$ is reaction rate, $f_i$ fugacities ($f_i = X_i \gamma_i P$ where $X_i$ is equilibrium concentration, $\gamma_i$ the activities coefficient, P system pressure), and $k_p{}^{\circ}$ is the equilibrium constant at standard conditions. Parameters $\alpha$, $\beta$, $\psi$ and $\phi$ are functions of temperature and the catalyst. These must be determined experimentally.

Uchida and Ogino[22] studied zinc oxide-chrome oxide catalysts over the pressure range 8.83-14.70 MPa (1280-2133 psi) and used a variation of the rate equation based on the Tempkin logarithmic adsorption isotherm. Using similar relations, Vlasenko et al.[23] examined the effect of temperature, pressure and volumetric flowrate on the methanol yield over an industrial zinc-chromium catalyst as a function of size. Higher temperatures were found to improve the yield, but the purity of the methanol formed decreased. Smaller particle size was found to improve both the yield and the quality. Increased pressure and volumetric flowrate also improved both yield and quality of the methanol formed.

Bakemeier et al.[24] attempted to describe the gross reaction rates experienced in industrial reactors. The investigation covered the water-gas shift reaction found in the inlet gas, in addition to two side reactions forming methane and methyl ether. Empirical correlations were developed.

Ferraris and Donati[25] analyzed the data of Natta et al. and applied a non-linear regression analysis in obtaining rate expressions. Several bi- and tri-molecular rate equations were studied and reduced to two biomolecular expressions.

Kafarov et al.[26] developed methematical models for the reactions based on low-temperature copper catalysts. Simulation studies on the computer were conducted to examine variations in the temperature, contact time, hydrogen:carbon monoxide ratio, and the carbon dioxide concentration on methanol yield for both laboratory- and commercial-scale reactors.

## INORGANIC SOURCES OF SYNTHESIS GAS FEEDSTOCKS

Although organic matter is the traditional source of raw material for synthesis gas feedstock, a large source of inorganics is available. These consist primarily of the separate oxidized forms of carbon and hydrogen, namely carbon dioxide, carbonates and water. The energy needed to reduce and combine these sources must be provided from other materials. By contrast, biomass or fossil fuels represent both elements and energy stored together. The use of inorganic sources in the production of synthesis gas is illustrated by the overall methanol cycle in Figure 3.4.

Carbon dioxide, found in plentiful quantities in the atmosphere, is the most readily accessible source of carbon. It exists, however, in a fairly dilute state (average concentration of about 0.33%). Atmospheric carbon dioxide is almost two orders of magnitude more than the annual combustion of fossil fuels.[27] Large quantities of carbon dioxide are also found dissolved in waterways.

The theoretical energy required to concentrate atmospheric carbon dioxide to one atmosphere as calculated by Steinberg et al.[28] is approximately 448 kJ/g-mol $CO_2$ (or 19.7 kJ/g-mol of $CO_2$) at $25°C$ and 1 atm. In comparison, 287 kJ/g-mol is required to concentrate oxygen.

Various methods are available to separate carbon dioxide from the atmosphere and concentrate it into a usable state. These methods include adsorption and stripping, and compression/refrigeration. In adsorption and stripping, $CO_2$ can be concentrated from liquids, such as water, caustic or carbonate solutions, or derived from solids via molecular sieves or lime.

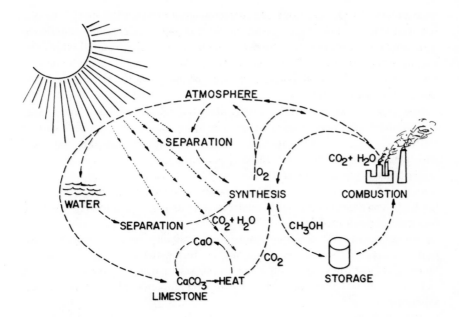

Figure 3.4  Methanol generation cycle from mechanical collection of $CO_2$ for synthesis gas feedstock.

Stripping seawater of dissolved carbon dioxide has been proposed as one specific approach.  Efficiencies, *i.e.*, yields, range from 21% for stripping to 0.8% for high-pressure water adsorption/stripping.  By contrast, separating oxygen from air by liquefaction is only about 10% efficient.  In many of these methods, water can also be obtained in quantities in excess of those required for electrolysis.  As such, for a given source of energy, both carbon dioxide and water can be obtained in sufficient quantities from the air quite readily.  The energy required to strip carbon dioxide from seawater or carbonate solutions is less than 10% of that required to electrolyze water; however, the capital investment is considerably greater.

Dissolved carbon dioxide found in the ocean is in equilibrium with the $CO_2$ in the atmosphere, and also with carbonates and bicarbonates dissolved and precipitated in the ocean.  The ocean can be viewed as a vast warehouse of carbon, which can be extracted by stripping seawater of carbon dioxide. Any significant quantity of $CO_2$ added to the atmosphere or extracted from the ocean will eventually be brought back into equilibrium by the huge oceanic carbonate reservoir.  The rate at which equilibrium is reestablished depends on the deep ocean transfer rates, which are relatively slow.  As such, local pulses of $CO_2$ additions or subtractions can occur in the short time of a few centuries, which might result in an imbalance in climatic conditions. This time lapse can, however, be overcome by extracting carbon dioxide from the air, either mechanically or via biomass, and returning it through combustion. This would avoid possible climatic changes resulting from increased atmospheric carbon dioxide.

Another source of carbon consists of terrestrial carbonates. The heating of limestone has been suggested as one approach to tapping this carbon source.[29-31] Large amounts of limestone are required with this approach, which creates a potentially serious environmental problem.

Carbonate solutions may be electrolytically reduced directly to formic acid and then to methanol. This is in contrast to extracting the carbon dioxide and then combining it with electrolytic hydrogen. At present, this technology is undeveloped.[32]

The oxygen required in methanol is also available from the inorganic carbon sources. It is also readily available during the electrolysis of water. When organic sources of carbon are employed, additional oxygen required can be obtained from air, either by restricted combustion or mechanical separation.

Hydrogen needed can be obtained from water by conventional hydrogen synthesis via electrolysis.[33] There are, however, other routes for the generation that can be taken, such as thermochemical or physical methods. Nuclear energy has also been proposed as a means of reducing water to hydrogen, or carbon dioxide to carbon monoxide.[34]

## ORGANIC SOURCES OF SYNTHESIS GAS FEEDSTOCKS

The conventional schemes for methanol production and its synthesis feed gas preparation have been entirely from sources of organic origin. Wood and other contemporary biomass were initially the major sources. The relative abundance of petroleum shifted sources to the inexpensive fluid fossil fuels. Biomass and the fossil fuels utilize the process of photosynthesis for collecting and storing solar energy. Plant life collects and reduces carbon dioxide and water from the environment and converts them to organic compounds. All the elements and energy are thus collected in one convenient form, which is used both as a feedstock and fuel. The feedstock can then be pyrolyzed to carbon monoxide and hydrogen. The gas is purified of sulfur compounds and then reacted with steam or carbon dioxide to approach the stoichiometric ratio of $2 H_2$ to 1 CO.

Figure 3.5 illustrates the relative magnitudes of various carbon resources. However, although large quantities are available, only a fraction of this can be practically recovered/harvested. The fluid fossil fuels have been used for the majority of the fuel and energy requirements of the industrialized countries. Based on current technology and energy costs, roughly 30% of the crude oil in the ground is considered recoverable. Of this amount, nearly half is estimated to already have been recovered in the U.S. and over 90% will be within the next three decades.[36] Similarly, the recoverable reserves of the entire world are projected to be exhausted within the next five to six decades. This rapid consumption of fossil fuels is forcing consideration of untapped sources of fossil fuels as well as biomass and sythetic fuels from inorganic sources.

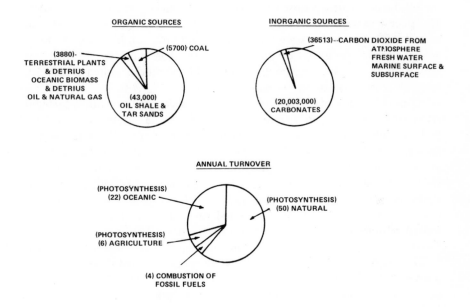

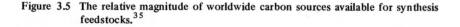

**Figure 3.5** The relative magnitude of worldwide carbon sources available for synthesis feedstocks.[35]

Coal is the most abundant fossil fuel after petroleum and natural gas. The basic resource is considerably larger, but subject to more stringent restrictions on recoverability and utilization. Recovery of a given coal deposit may range from 25-90%; however, only 8% of the entire U.S. coal resources is considered recoverable.[35] Environmental restrictions limit the amount of this coal that can be burned or extracted. Coal with large amounts of sulfur can still be used in chemical plants, but additional processing and expense are required for sulfur removal.

Coal has been used as a feedstock in methanol production. Naphtha and natural gas, however, replaced its importance in methanol production. Coal gasification for methanol production is technologically feasible and there are a number of developmental programs in progress. Investigations are being conducted to examine plant design variations involving the use of different types of coal, pressures, heat transfer methods, use of air or oxygen, and feed mechanisms.[37-39] In general, coal gasification appears competitive with other synthetic fuels. Large-scale production of methanol from coal is likely to proceed as soon as there is a guaranteed demand and price.

Oil shale and tar sands represent the largest fossil carbon resources. Recovering oil from shale involves crushing the shale and heating it to release the oil. Large quantities of shale are necessary to make this approach feasible. The richest shale contains only about 5-10% oil by weight. Design experimental and pilot systems are under study, but costs are high and the technology undeveloped. Actual utilization of oil shale will depend highly on resolution of associated environmental difficulties and the limited availability and high costs of alternative resources.

As noted earlier, wood supplied a major portion of the world's fuel and chemical feedstocks about a century ago. Wood has often suffered from limited availability when it was considered a finite, rather than a renewable, resource. Attention is now being given to biomass as a fuel and feedstock on a renewable basis. Biomass quantities are subject to variation with the weather; however, comparative orders of magnitude are sufficient to demonstrate its large potential as a fuel or feedstock. As will be discussed later, the productivity of much of the natural vegetation and forests can be improved significantly through cultivation and proper management. Managed forests, for example, are four to six times more productive than virgin forests.

Large-scale usage of biomass for energy production is likely to compete heavily with the primary demands for food and fiber. Traditionally, agriculturally suitable land has had grain production as a primary focus.

Urban waste is another source of biomass and, although small in comparison to the other sources previously discussed, can make a significant energy contribution. The per capita production of municipal waste in the U.S. is on the order of 1.5 kg/day. This means that a city of 1 million would generate 550,000 tonne/yr of waste. Roughly 25% of this refuse is carbon and another 25% water. The Purox® process developed by Union Carbide can gasify solid wastes. Design studies have been conducted to examine the prospects of converting municipal solid waste to methanol or ammonia using the Purox process.[40-48] At present, the economics favor the production of ammonia over methanol.

Supplies of carbon monoxide and carbon dioxide can also be obtained from secondary sources. For example, there are a number of manufacturing operations, such as steel mills and power plants, which produce large quantities of carbon monoxide that could be used in methanol synthesis.[49] Conceivably, carbon dioxide generated from power plants could be collected and recycled through methanol plants.

## 4

## METHANOL SYNTHESIS FROM WOOD WASTES

### APPLICATION OF GASIFIERS

Basically, any carbonaceous material, such as coal, lignite, wood waste, agricultural residue and municipal refuse, can be used in the production of synthetic methanol. By contrast to natural gas, these materials require several additional processing steps to refine the crude synthesis gas into syngas. The final gas product consists of two parts of $H_2$ to one part of CO. Because of these additional steps, the conversion of a carbonaceous material is more energy-intensive than that required by natural gas. In addition, its yield is considerably less. Figure 4.1 illustrates the details of the process.

According to Figure 4.1, the carbonaceous material is first partially burned or oxidized, producing a crude gas consisting mainly of $H_2$, CO and $CO_2$. If air is used to oxidize the feed material, the gas consists of roughly 46% nitrogen, which can be removed by cryogenic means. If oxygen is used, a cryogenic system is necessary for initial separation of air into oxygen and nitrogen.

Several designs for gasifiers have been devised, capable of partially oxidizing wood, wood waste and refuse. These systems are designed to operate at atmospheric pressure. By contrast, coal gasifiers are designed to operate at pressures up to 400 lb/in.$^2$ g. Gasifiers produce a crude gas consisting primarily of $H_2$, CO and $CO_2$, with minor amounts of heavier hydrocarbons. In addition, about 2% of the wood on a dry basis is transformed to an oil-tar fraction.

Two commercial systems that have received considerable attention are the Purox® system, developed by Union Carbide, and the Moore-Canada gasifier. The Purox system uses municipal refuse as its feedstock. It is believed, however, that the crude gas composition is essentially the same for wood waste because garbage has been found to have practically the same composition with regard to carbon, hydrogen and oxygen.[50]

The Purox reactor-gasifier uses a moving bed, in which oxygen is passed countercurrent to the downflowing residue. As the material flows through

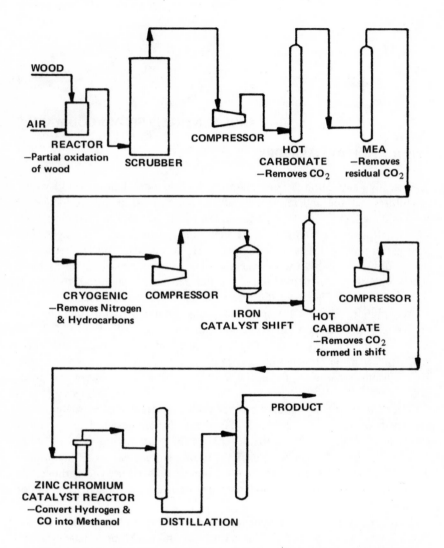

Figure 4.1. Details of the process for methanol synthesis from wood residues.

the reactor, it passes through various stages of drying, reduction and oxidation. At the bottom of the unit, ash is removed in a molten state at a temperature of about 1650°C. Crude gas containing a large amount of moisture leaves the top at a temperature of about 93°C.[50,51]

In 1970, Union Carbide initiated developmental work in a pilot plant sized for a feed rate of 2.5 oven dry tons per day (odt/d) of municipal refuse. The first demonstration plant was put into operation in 1974 and consisted of a 3.05-meter diameter reactor, having a capacity of 150 odt/d of garbage. Scaleup of the reactor to this level resulted in a significant decrease of conversion efficiency. This was evidenced by an increase in carbon dioxide and hydrocarbon levels in the raw gas.

The stability of the bed plays an important role in the performance of any moving bed reactor. Stability apparently decreases with increasing diameter in this system. The diameter limitation has been demonstrated in the gasification of coal by the Lurgi Process. Lurgi succeeded in operating a reactor 3.57 meters in diameter. It was not possible, however, to put into operation a reactor 4.2 meters in diameter.

Moore-Canada of Richmond, British Columbia, developed a moving bed reactor for producing a gas of low heating value from "as-is" wood waste. In contrast to Purox, the Moore reactor uses air as the oxidizing medium. Because of the high nitrogen content, the raw gas has a heating value of about 180 Btu/scf, in contrast to that of the heating value of the Purox unit of 350 Btu/scf.

Currently, a semiworks unit is in operation, consisting of a 1.68-meter-diameter gasifier with a capacity of about 18 odt/d of wood waste.

The Moore reactor's operation is similar to Purox in that the feed material enters at the top, and the wood ash is discharged from the bottom. However, because air rather than oxygen is used, the maximum temperature of the oxidation (lower) zone is only about 1204° C. Waste material is discharged as a solid in a granular form rather than as a molten slag. Pressure at the base of the reactor is approximately 6-8 lb/in.$^2$g and, at the top 2-3 lb/in.$^2$g. The raw gas leaves the reactor at a temperature between 71°C and 82°C. By purging the air with steam, the hydrogen content of the crude gas from the Moore reactor-gasifier increases from 8-10% to 18-22%.

There are a variety of other gasifier designs. These are summarized below:

*Battelle* – The Battelle Pacific Northwest Laboratories carried out a 1-year pilot-plant study on the partial oxidation of municipal refuse in a 0.91-meter moving bed reactor. This study included partial oxidation of wood chips using air and steam, in which similar results to those reported by Moore-Canada were obtained.[52]

*Thermex* – Alberta Industrial Development, Ltd. has put into operation a 50-odt/d demonstration plant for the gasification of wood waste. In its present operation, the system generates a char and low-Btu gas; however, it can be designed to operate without forming char. The gasifier is a fluidized-bed design. The wood waste feed must be hammermilled to less than 5.08 cm particle size.

*Copeland* -- The Copeland organization has designed and constructed several fluidized-bed reactors for the pulp industry for the disposal of the organic matter in waste liquor. These units are capable of accepting "as-is" wood waste and sludge, but their applicability to partial oxidation for syngas has not been investigated.

*Lurgi* -- The Lurgi reactor is designed to gasify coal with oxygen and steam at 300-400 lb/in.$^2$g. It can handle only noncaking-type coal, with particle size ranging from 0.95-5.08 cm. No attempt has been made to process wood waste in a Lurgi reactor. The reactor requires relatively uniform particle size and is not to be expected to handle wood waste.

*Winkler* -- The Winkler system is also a fluidized-bed coal gasifier operating at or near atmospheric pressure. These units are typically 5.49-meter in diameter and operate at a temperature of about 1204 °C. Coal fed to the unit is ground to less than 0.635 cm. No attempt has been made to apply the gasifier to wood waste. Since particle size is limiting, it is doubtful that it could be applied to handling wood wastes.

*Koppers-Totzek* -- These units process pulverized coal with steam and oxygen under slagging conditions at atmospheric pressure, with temperatures up to 1927 °C. Although a number of Koppers-Totzek installations have been built, this type gasifier is not viewed as being practical for the handling of wood waste because of requirements for finely ground feed.

## PURIFICATION OF THE SYNTHESIS GAS

Crude gas from partial oxidation units must be processed to remove water vapor, tars, organics, hydrocarbons, nitrogen and $CO_2$. The clean gas consists primarily of $H_2$ and CO. The gas is processed in a shift reactor to react part of the CO to form additional $H_2$, so the final gas contains the proper ratio of two parts $H_2$ to one part CO. In the shift reactor, additional $CO_2$ is formed, thus making it necessary to again scrub the gas prior to entering the synthesis reactor.

The unrefined gas exiting the gasifiers passes upward through a single cooler-absorber-scrubber. During this step the gas is cooled from about 82 to 32°C in three stages of contacting (Figure 4.1). In the lower two stages, the crude gas is contacted by cooled, recirculated liquor streams. In the upper stage, reclaimed water condensate is used to complete the removal of organic compounds, such as acetic acid.

Moisture is condensed from the crude gas in an amount roughly equal in weight to the dry wood substance entering the system. The water contains about 2% of the soluble organics, and it is necessary to clean the stream for environmental purposes. Organics recovery can be achieved with a suitable solvent, such as methyl ethyl ketone in a liquid-liquid multiple-stage extraction operation. The extract, or light-density phase, is processed in an extraction tower to recover the solvent overhead, and the organic-rich material from the bottom. The heavy density raffinate phase is processed in a raffinate stripper, which recovers that portion of the solvent dissolving in the water phase. From the bottom of the raffinate stripper, the effluent is essentially a water product of low biochemical oxygen demand (BOD). The organic product obtained from the extraction tower has been suggested for use as a fuel in the boiler. It may also be economically feasible to separate the stream into its components, mainly acetic acid, for by-product value.

The cooled and partially purified gas is then compressed to about 100 lb/in.$^2$ g and treated in a two-stage system to remove carbon dioxide. In the first stage, a hot potassium carbonate system is used to reduce $CO_2$ content to about 300 ppm. As shown in Figure 4.1, concentration is reduced to about 50 ppm, using monoethanolamine as a scrubbing agent.

The clean compressed gas passes to a cryogenic system, and in a series of switching exchangers, the residual $CO_2$ and water vapor are removed to prevent freezeup in the downstream exchangers and distillation towers. Methane and hydrocarbons are then removed. Cryogenic distillation is used to separate CO from nitrogen. The liquid nitrogen exiting the system is used to precool the incoming gas. The purified product gas consists of a mixture of carbon monoxide and hydrogen. The product gas requires further processing as, at this point, it is not in the ratio of 2:1 of $H_2$ and CO required for syngas (which is the proper stoichiometric ratio necessary to produce methanol). The so-called "shift reaction" represents the final purification stage.

Following the cryogenic separation of the inerts, the gas is compressed to 400 lb/in.$^2$ g for the shift conversion. Part of the CO reacts with water vapor in the presence of an iron catalyst to form additional hydrogen to the extent that the final gas contains the required two parts hydrogen to one part carbon monoxide. Equation 4.1 describes the reaction, which is exothermic in nature.

$$CO + H_2O \xrightarrow{\text{catalyst}} H_2 + CO_2 \qquad (4.1)$$

The hot potassium carbonate absorption system removes about 97% of the carbon dioxide formed during the shift reaction. The SNG is then compressed to a pressure ranging from 1500-4000 lb/in.$^2$ g and fed into the methanol synthesis reactor. In the reactor, approximately 95% of the gas is converted to methanol. The remainder passes as inerts to the boiler.

There are basically two processes for methanol synthesis: (1) the Vulcan process, which uses a zinc-chrome catalyst operating at pressures ranging from about 2000-4000 lb/in.$^2$ g and (2) the ICI copper catalyst process, operating at pressures ranging from 1000-2000 lb/in.$^2$ g. A major consideration in establishing the system pressure for either process is the purity of the feed. With increasing amounts of impurities, the system requires higher pressure to minimize the concentration of inerts leaving the system. The crude methanol product from the synthesis reactor is then sent to a distillation train for separation of the light ends and higher alcohols from the methanol product. The mixture of light ends and higher alcohols is used as a fuel in the boiler.

## ECONOMIC FACTORS AND PLANT CAPACITIES

Production costs depend largely on capital investment and raw materials costs. One approach to reducing unit costs is to build a high-capacity facility because investment for scaleup generally increases with production by a 0.6 factor.[50] The trend in the synthetic methanol industry has been to increase plant sizes from about 50 million gal/yr to 200 million gal/yr. There are a dozen plants in the U.S. that can produce 1.2 billion gal/yr of methanol (Table 4.1).

Table 4.1  U.S. Methanol Production[50]

| Number of Plants | Total (million gal/yr) |
|:---:|:---:|
| 1 | 22 |
| 3 | 150 |
| 1 | 80 |
| 4 | 400 |
| 1 | 160 |
| 1 | 200 |
| 1 | 230 |
| Total U.S. Capacity | 1242 |

Hokanson and Rowell[50] estimated the operating costs for a 50 million gal/yr synthetic methanol plant from wood waste and determined the effect of scaleup on cost-preparing estimates for a facility capable of producing 200 million gal/yr. They compared their estimates with figures of investment and operating costs for facilities using natural gas and coal as raw materials. The analysis was based on facilities employing boilers to produce steam to generate electricity and to drive turbines required for compression. As such, the plants were considered to be self-sufficient, *i.e.,* requiring no outside utilities other than cooling water makeup.

The investment requirement estimated for a 50 million gal/yr methanol plant using wood waste was $64 million, based on 1975 economics. A breakdown of these estimates is illustrated in Figure 4.2. Note that the breakdown includes offsite utilities, wood yard handling facilities, finished product storage, and office and laboratory buildings. Hoakanson's estimates include a

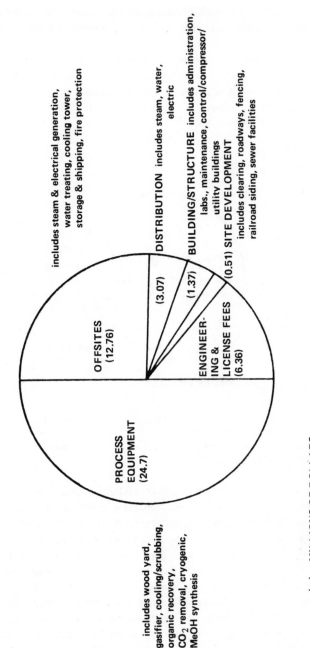

Figure 4.2 The investment estimate for a wood waste methanol plant with a capacity of 50 million gal/yr.[50]

contingency of 25% and a working capital of 5%; however, no provisions were made for the expected escalation in cost of equipment and construction labor.

Estimated operating costs for the 50 million gpy methanol plant are illustrated in Figure 4.3. Estimates on production costs include fixed costs, raw material, labor and overhead. Fixed costs are based on an allowance of 8% for depreciation, 4% for maintenance (including labor and material), and 2% for local taxes and insurance.[50]

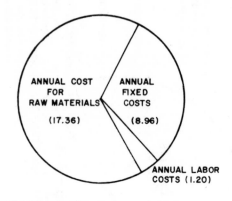

( ) - MILLIONS OF DOLLARS

Figure 4.3   Breakdown of operating costs for a wood waste methanol plant with a capacity of 50 million gpy.[50]

The investment cost estimates made by Hokanson and Rowell[50] represent roughly three times the cost for a comparable plant based on a natural gas conversion to methanol. Also, the conversion efficiency of natural gas to methanol is significantly greater than that of wood waste. A comparison of efficiency of conversion of natural gas, coal and wood waste is given in Figure 4.4. (Note that the plant efficiency is based on the heating value of methanol as a fraction of the total energy input into the plant.) In the case of coal conversion to methanol, although considerably more efficient than that of waste wood, a conversion plant involves more processing facilities because of the greater amount of ash and sulfur (wood has no sulfur). Coal conversion to syngas is more efficient mainly because it has a higher carbon content and less oxygen than wood.

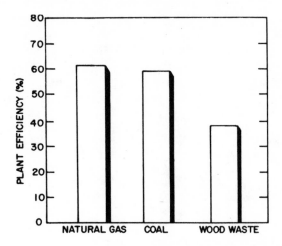

Figure 4.4  Methanol synthesis conversion efficiencies for natural gas, coal and wood waste.[50]

## 5

## AUTOMOTIVE USES OF METHANOL

### PROPERTIES OF METHANOL-GASOLINE BLENDS

There is considerable disagreement about the effects of methanol in automobile engines. Some basic properties of isooctane and methanol are compared in Table 5.1. To achieve the ideal chemical combustion of gasoline, *i.e.,* in the proper stoichiometric ratio, an air:fuel ratio of about 14.5 is needed. This means that 14.5 parts of air to 1 part of gasoline is theoretically required to achieve the proper combination for complete combustion of the fuel. For methanol, the air:fuel ratio is only about 6.4. As a result, a car that has its carburetion set for gasoline will undergo automatic leaning if methanol is added. That is, even without touching the carburetor settings, the air:fuel ratio will be excessive. An engine that is running on the lean side usually has fewer emissions and better gas mileage; however, a drop in performance can be expected. With current emission control standards, newer cars take as much advantage of this effect on emissions as performance will allow. Also, they have their carburetors set as lean as possible.

The addition of methanol to gasoline and no readjustment of the carburetor can result in some reduction of emissions. Gas mileage is, however, another factor that must be considered. The heating capacity of gasoline is roughly double that of methanol. This means that on a Btu basis, it takes roughly twice the quantity of methanol to drive the same distance as it does with gasoline. The addition of methanol to gasoline, without a carburetion change and with the leaning effect (which should give better mileage), becomes complicated with the lower Btu content of the fuel. The net result is a reduction in mileage.

### EXPERIMENTAL INVESTIGATIONS

Reed and Lerner[53] and others[54,55] have performed actual road tests on cars using methanol-gasoline blends. In testing unmodified cars over a fixed course using methanol-gasoline blends, it was found that mixtures between

Table 5.1 A Comparison Between the Properties of Isooctane and Methanol

| | Isooctane $(C_8H_{18})$ | Methanol $(CH_3OH)$ |
|---|---|---|
| Molecular Weight | 114.224 | 32.042 |
| Carbon:Hydrogen Weight Ratio | 5.25 | 3.0 |
| Carbon, % by weight | 84.0 | 37.5 |
| Hydrogen, % by weight | 16.0 | 12.5 |
| Oxygen, % by weight | 0.0 | 50.0 |
| Boiling Point, °C at 1 atm | 99.24 | 64.5 |
| Freezing Point, °C at 1 atm | -107.4 | -97.8 |
| Vapor Pressure, psia at 37.8 °C | 1.708 | 4.6 |
| Density, 15.5 °C, lb/gal | 5.795 | 6.637 |
| Coefficient of Expansion 1/F at 60 °F and 1 atm | 0.00065 | 0.00065 |
| Surface Tension, dynes/cm at 20 °C and 1 atm | 18.77 | 22.61 |
| Viscosity, centipoises at 20 °C and 1 atm | 0.503 | 0.596 |
| Specific Heat of Liquid, Btu/lb - °F at 77 °F and 1 atm | 0.5 | 0.6 |
| Heat of Vaporization, Btu/lb at boiling point and 1 atm | 116.69 | 473.0 |
| Heat of Vaporization, Btu/lb at 25 °C and 1 atm | 132 | 503.3 |
| Heat of Combustion, at 25 °C | | |
| Higher heating value, Btu/lb | 20,566 | 9,776 |
| Lower heating value, Btu/lb | 19,065 | 8,593 |
| Lower heating value, Btu/gal | 110,480 | 57,030 |
| Stoichiometric Mixture, lb air/lb | 15.13 | 6.463 |
| Octane Number (research) | 100 | 106 |

5 and 15% increased fuel economy and performance and lowered carbon monoxide emissions and exhaust temperatures.[55] Experiments also showed an elimination of knock on one engine. Reed and Lerner[55] attributed the improved methanol performance to two factors: chemical leaning and the dissociation of methanol near 200°C. At about 200°C, energy can be absorbed during the compression stroke of the engine, accompanied by a release of up to 40% hydrogen for a 10% mixture. Figure 5.1 shows a sampling of Reed and Lerner's data on performance for several car models.

Santa Clara University and the city of Santa Clara have performed methanol fuel tests on city vehicles.[56] Results obtained were similar to those of Reed et al.[55] showing an overall lowering of emissions and improved mileage.

Breisacher and Nichols[57] performed tests of methanol blends on a Wiscon engine. The study concluded that exhaust emissions on the whole

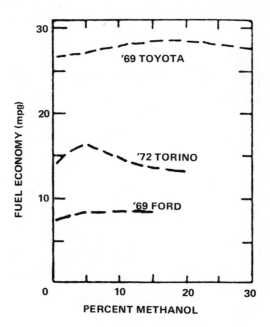

Figure 5.1  Car performance data by Reed *et al.*[55]

were not degraded by the addition of methanol; however, carbon monoxide levels were observed to decrease as the quantities of methanol in gasoline were increased. The investigators attributed this latter effect to leaning. Some of the emissions data obtained from the study are shown in Figure 5.2. The study was extended to several field tests of methanol-gasoline blends. In each case elimination of knock was made possible by the addition of the methanol to the gasoline.

The Institute of Gas Technology (IGT), General Motors Corporation (GMC), Chevron Oil Company and the Bureau of Mines each have tested methanol-gasoline blends. Fuel economy, *i.e.,* mileage on the cars that were tested, was observed to drop. In four cars tested by the Bureau of Mines, using both city and highway driving conditions, preliminary results indicated that in every case, the addition of methanol to gasoline, although usually lowering the miles per gallon, increased the miles per energy unit consumed. As such, the cars showed better efficiency on a miles per Btu basis with the methanol-gasoline blends.

The emissions results obtained by General Motors[58] agreed with the work of Breisacher and Nichols. Nine cars were tested in the GMC study with

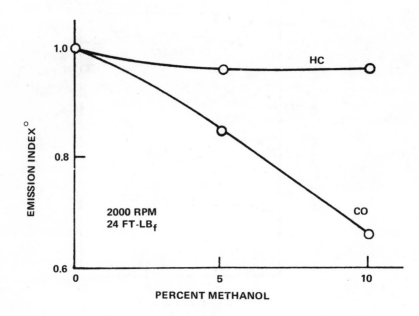

Figure 5.2   Hydrocarbons and CO emissions measured by Breisacher and Nichols[57] for several methanol-gasoline blends on a Wisconsin engine.

existing carburetion and spark timing.  Comparison was made between unleaded gasoline and a 10% methanol-gasoline blend, revealing an overall reduction in CO but no consistent and overall change in the $NO_x$ and hydrocarbons emissions.

The antiknock property or octane quality of methanol may be one of its most important advantages.  Ingamells and Lindquist[59] noted that methanol's most desirable feature as a motor fuel is its high octane quality. Methanol rates 106-115 octane numbers (ON) by the research octane method (laboratory) and 88-92 ON by the motor octane method (laboratory). Tests indicate methanol to be equivalent to the highest octane gasoline based on the motor test and much better based on the research method.  Of the two test methods, the motor test is considered more severe because it often results in lower octane ratings.

Ingamells and Lindquist[59] also noted methanol's ability to "boost" gasoline's octane quality.  In this study, a comparison was made of average data obtained with three different unleaded gasolines. The research ON was observed to increase by 4.4 ON by the addition of 10% methanol, resulting in an extremely high blending value of 135 ON. (Note that the blending value is computed from the measured octane data using a linear expression.) The addition of 10% methanol resulted in increasing the motor ON by 2.0 ON, resulting in a blending value of 102.5. Chassis dynamometer tests were conducted on a number of cars to provide road octane data on the same fuels. Results were obtained in a range between the research and motor ON data, which is normal for motor gasoline.[59]

Although antiknock characteristics associated with methanol blending agents in unleaded gasoline are almost universally accepted, certain key problems must be considered before methanol usage on a large-scale basis can be recommended. Some of these problems are discussed in the following section.

## PROBLEMS ASSOCIATED WITH METHANOL-GASOLINE BLENDS

Reed et al.[55] noted that problems arose with hesitation on warmup with the blended fuel. Problems with hesitation were reported to increase with higher percentages of methanol addition to fuel.

In the Chevron study, drivers made a daily record using a demerit basis for rating the performance deficiencies that occurred during a commuter trip. The demerit system was based on the number of malfunction occurrences and the severity of each. Primary deficiencies noted included stalling, hesitation on opening of the throttle and surging under some conditions. Figure 5.3 gives comparisons of the average number of demerits received from gasoline and from a 10% methanol-in-gasoline blend. Chevron established a maximum acceptable drivability level of five demerits.

Deterioration in performance was attributed to the following:

1. excessive leaning effect, caused by the addition of the methanol;
2. separation of the methanol and gasoline in the fuel system, due to the absorption of water into the fuel mixture; and
3. deterioration of existing fuel system components, due to corrosion and/or electrolysis.

The third problem appears to be due mainly to the fact that alcohols are basically good solvents. Chevron has noted problems associated with the decomposition of various portions of the cars' fuel systems and the eventual clogging of filters. Through the use of different types of fuel tanks and alternate materials throughout the fuel system, this problem could be eliminated.

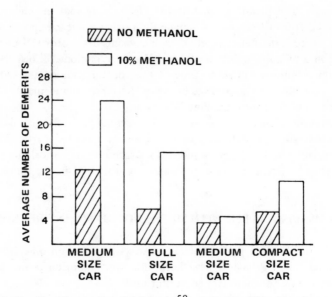

Figure 5.3  Results of Ingamells and Lindquist[59] on the drivability of various model cars using 10% methanol-gasoline blend.

Absorption of water into the fuel mixture appear to be the most serious problem associated with methanol-gasoline blends in present cars. Although it is feasible for gasoline and anhydrous methanol to be mixed at the fuel pump, the presence of water in the air, the strong attraction of alcohol to water, and the inability of methanol-gasoline solutions to accept much water before phase separation occurs, suggests that methanol and gasoline may separate within the fuel system. Stratification will occur, with methanol and water at the bottom of the vessel and gasoline at the top. As noted earlier, an engine set to operate on gasoline will not perform with 100% methanol. The separation of the methanol to the bottom of the carburetor bowl will stall the engine, or it simply will not start until the gasoline portion of the separated fuel is finally sent to the cylinders.

Wigg[60] has noted one further problem, relating to vapor lock. Even with the addition of small percentages of methanol to gasoline, the increase in the Reid Vapor Pressure (RVP) is substantial (Figure 5.4). The RVP increases in some cases exceed some state vapor pressure limitations for gasoline. According to Wigg, this increase in Reid Vapor Pressure and the corresponding decrease in the boiling point of the methanol-gasoline blend result in a vapor lock problem.

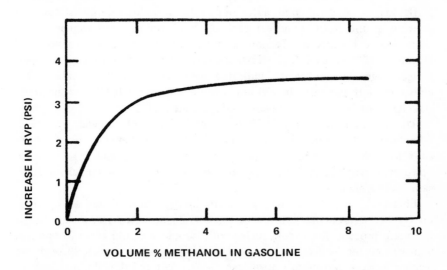

Figure 5.4 The effect of methanol addition to gasoline on vapor pressure.[60]

The control of vapor pressure of a methanol-gasoline blend might be accomplished by removing some of the butane and propane content of the gasoline. This would, however, have negative effects on mileage and octane properties of fuel.

As shown by various investigators, there are some real advantages in the addition of methanol to gasoline, especially with the excellent octant properties of the methanol. However, the significant problem areas noted must be further investigated and solved before wide-scale promotion of methanol-gasoline blends can be recommended.

## PURE METHANOL AS A FUEL

The main application of 100% methanol as a fuel for the automobile has been limited to racing cars. It has been known for some time that methanol developed more power than gasoline and that compression ratios as high as 15:1 were possible with methanol fuel.[61]

Starkman *et al.*[62] described performance comparisons of methanol, ethanol, benzene and isooctane as tested on a CFR test engine. Isooctane is often used as a pure chemical representative of gasoline. At stoichiometric equivalence ratios, methanol develops slightly more power than isooctane. At very rich equivalence ratios, methanol develops significantly more power than isooctane. Starkman *et al.*[62] showed that with this increased power, methanol gives a significantly greater fuel consumption than isooctane.

Because of concerns about automobile exhaust pollution, interest has been growing in the emissions effect of 100% methanol-powered vehicles. The Consolidated Engineering Technology Corporation retrofitted a car to operate on 100% methanol fuel. Tests on this vehicle and also on a methanol-fueled single-cylinder engine indicated decreases in nitrous oxide ($NO_x$) emissions, but increases in alcohol, carbon monoxide (CO) and aldehyde (HCHO) emissions when compared with gasoline fuel.

The Ford Motor Company[63] has tested blends of toluene, isooctane, *n*-heptane, and up to 25% methanol on a single-cylinder engine. Ford's study concluded that methanol has little effect on both $NO_x$ and CO emissions; however, it was noted that at rich equivalence ratios, HC and aldehyde emissions were higher.

Ebersole[64] noted that the lean misfire limits with methanol are approximately two equivalence ratios leaner than with isooctane. This means that in a leaning process, the engine fueled with isooctane would cease to operate, while the engine fueled with methanol would continue to operate through the leaning process two more equivalence ratios. As a result, unburned fuels in the exhaust were much less, zero CO emissions were maintained, and nitrous oxides were slightly less; however, aldehyde emissions were substantially greater when comparing methanol to isooctane.

Potential air emissions problems from the use of 100% methanol as a fuel are not fully understood. Ebersole[64] and others note that amounts of aldehydes and unburned methanol appear in the exhaust, especially with leaner than stoichiometric air:fuel ratios. The overall environmental effects of widespread amounts of these residues simply are not known.

There are a number of serious problems associated with modifying an engine to run on 100% methanol. With the much lower air:fuel ratios that are required, a different carburetion or fuel injection system is needed to accommodate the large quantities of fuel that must be metered to the engine. It is generally believed that a fuel injection system is appropriate. Higher compressions are suggested to take advantage of the very high octane values of pure methanol. An induction heating system is, however, necessary to vaporize the fuel.

Another major obstacle is starting engines in cold weather. It has been reported that methanol-fueled vehicles are virtually impossible to start at low temperatures, due to the very high latent heat of vaporization of the fuel. Racing cars using methanol blends as a fuel are started either by very high speed starters or by pushing.

In summary, it appears that the following conclusions can be made concerning the use of 100% methanol as a fuel for automobile engines:

1. Engine modifications may be needed to include larger carburetor jets of fuel injection and some type of intake manifold heating to obtain methanol vaporization.
2. Satisfactory performance has been reported at leaner equivalence ratios than with gasoline, resulting in lower $NO_x$ emissions.
3. Compression ratios on methanol-powered engines can be raised significantly to obtain more efficient fuel use. However, this may be accompanied by an increase in $NO_x$ emissions.
4. Exhaust emissions contain higher quantities of unburned methanol and aldehydes than with gasoline.
5. Cold starting problems occur because of the high latent heat of vaporization of methanol.
6. Fuel mileage can be expected to be half that of gasoline, implying a need for larger fuel tanks.
7. Fuel systems will require special construction materials to withstand the strong solvent properties of methanol.

Considerable interest has been shown in the U.S. in the development of an engine to optimize the combinations of power, efficiency and emissions control using 100% methanol as fuel. The benefits of such an engine for use in fleet automobiles in high population density and pollution areas are significant.

It is clear that a great deal of research on the subject of methanol-fueled vehicles is still needed. Primary emphasis should be placed on the continued study of emissions, especially aldehyde and methanol air pollution. Further study and engineering of optimized methanol-burning engines are necessary.

**6**

ETHANOL SYNTHESIS

## EARLY DEVELOPMENTS AND PRESENT WORLDWIDE
## ACTIVITIES

The knowledge needed to produce alcohol from grains and fruits dates back to ancient times. The Egyptians and Mesopotamians recorded methods to brew beer as early as 2500 B.C. Despite this ancient knowledge of converting sugar and starches to ethyl alcohol, modern techniques were developed only as recently as the middle of the 19th century. A German botanist, F. T. Kutzing, and a French chemist, Louis Pasteur, demonstrated that fermentation was attributed to the action of yeast cells, which generate enzymes necessary to convert sugars into ethanol. It was shown later that the complete yeast cell is not required for fermentation to take place, but that an appropriately prepared extract will suffice. It has also been demonstrated that certain acids and enzymes could be used to convert the complex nonfermentable cellulose molecule into the fermentable glucose molecule through hydrolysis. Glucose could then be converted into alcohol via fermentation.

The earliest systematic studies of ethanol production through the hydrolysis of cellulose were conducted in Germany during World War I. Two commercial processes developed at that time were the weak acid method (Scholler) and the strong acid method (Rheinau). The weak acid method was perfected in the United States and became the basis for the commercial development of the synthetic technique of ethanol production from petroleum-derived ethylene.

Historically, ethanol's primary use has been as a beverage. Before the discovery of petroleum in the last century, ethanol was also widely used for cooking, heating and lighting. Between 1935 and 1937, Henry Ford sponsored three conferences in Detroit on the industrial uses of such farm products as grains, soybeans and peanuts. The Model A was usually equipped with an adjustable carburetor, which was designed to allow combustion of alcohol, gasoline or any mixture of the two.

Until about the mid-1940s, the production of industrial ethanol from agricultural crops was widespread in Europe.  The main crops employed in ethanol production were corn, potatoes and sugar beets.  After 1945, crop fermentation for ethanol production in the industrial chemical market was replaced by petroleum-based sources because of lower production costs. Between 1920 and 1940 there was extensive experimentation with, and use of, a variety of substitute fuels in Europe. Alcohol blends were used with reasonable success in more than four million vehicles during this time period. Alcohol-gasoline blends were generally in mixtures of up to 25%.

During the post World War I era in England, there was a clause inserted in the Finance Act of 1920 legalizing the use of ethyl alcohol for fuel purposes. As a fuel, ethanol was free of all duty and restrictions in England.

Countries that used ethanol blends in gasoline during the 1920s and 1930s included Argentina, Australia, Cuba, Japan, New Zealand, the Philippines, South Africa and Sweden.

Heavy costs were incurred in European government-sponsored programs for the development of an ethanol production industry. Imported oil during this time sold for about 9¢/gal. However, the cost of alcohol averaged about 44¢/gal.  Although economically unattractive, programs directed at large-scale use of ethanol were justified because:

1.  countries wanted to achieve independence from imported petroleum in the event of war; and
2.  countries wanted to establish a secure source of ethanol for munition industry requirements, added benefits being the stimulation of home agriculture, labor and industry during the depression years, and reduced national trade deficits.

During the 1930s, the major interest in this country was in the foreign export markets.  Chrysler Motor Corporation produced cars that were modified slightly to accommodate shipments to New Zealand, which, at the time, was using 100% alcohol fuel.  Another example is International Harvester, which made trucks powered with engines designed to burn ethanol for export to the Philippines.

During World War II, Germany fueled most of its vehicles on alcohol derived from potatoes.  The U.S. constructed an ethanol plant in Omaha, Nebraska to produce motor fuel for the army.  In addition, to meet wartime needs, the U.S. recognized the potential for petrochemical products from grain-derived ethanol in the large-scale synthetic rubber production. By 1944, the biomass-produced alcohol industry had registered a sixfold increase in production over a period of five years.

During the postwar years, the cheap availability of petroleum gradually returned to the world.  Governments withdrew their subsidies to national ethanol production programs and, by 1950, nearly all facilities were terminated. Interest in the use of surplus or distressed grain for alcohol production has continued in the Midwest mainly because such a program would benefit farmers. In 1971, the Nebraska legislature passed several bills that established programs to assist in the development of the grain alcohol industry. An automotive fuel called Gasohol was developed and is being tested. Gasohol consisted of a blend of 10% agriculturally derived ethyl alcohol and 90% unleaded gasoline.

A number of nations are examining the feasibility of converting local crops into energy. Australia is actively investigating the development and manufacture potential of ethanol from eucalyptus trees. Brazil has implemented the largest program to use biomass feedstocks for the production of alcohol fuel.

Brazil imports more than 80% of the petroleum it requires and, as such, has been acutely affected by increases in petroleum prices in the past few years. Brazil's foreign trade deficit has risen from $17 billion to $31 billion in just four years (petroleum is imported at an annual cost of $4 billion).[65,66] This critical economic situation has provided incentive for the Brazilian government, as well as private industry, to embark on an ambitious program to develop a domestic energy resource based on the use of biomass.

Brazil is the fifth largest country in the world and the largest producer of sugarcane. The sugar crop totalled over three million tons in 1977. Roughly one million tons of this was processed directly into ethanol. In addition, most of the molasses produced from sugar operations was processed into alcohol.

In 1975, Brazil established the National Alcohol Commission to coordinate a massive effort to develop biomass-based fuels. The Commission decreed that standard fuels must be supplemented and eventually replaced by alcohol fermented from sugarcane, manioc, sweet potatoes and/or other starchy crops. Since the program's initiation, 141 fermentation and distillery projects have been authorized, with numerous others in the works. Newer facilities are being designed to burn bagasse, *i.e.,* sugarcane stalks and residues, as fuel for process heat in a further attempt to conserve petroleum.

A typical energy balance for an ethanol-producing plant from sugarcane is illustrated in Figure 6.1. In the diagram, the following inputs and outputs are identified.

1. energy input for growth, harvesting and collecting biomass feedstocks;
2. energy input for conversion processes;
3. energy output from the end product, ethanol; and
4. bagasse or Stover, which is the waste product.

Bagasse is reused as a primary energy input to the ethanol production process.

A production quota of 2 billion liters of anhydrous alcohol for fuel per year is the government's goal by 1985. It is estimated that 2% of Brazil's land area will be supplying up to 75% of its motor fuel by the year 2000.[67]

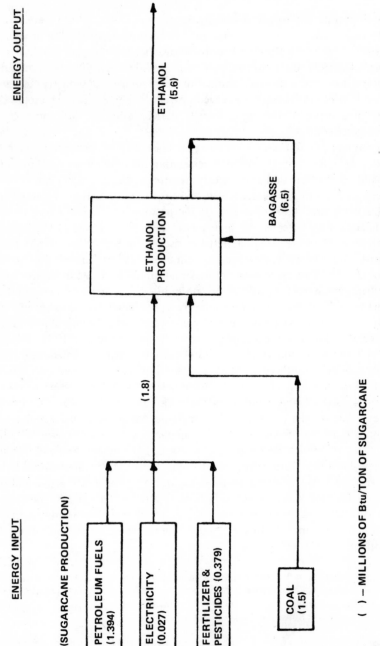

Figure 6.1  The energy balance for the production of ethanol from sugarcane.[66]

## LARGE-SCALE USAGE IN BRAZIL

Scores of government-owned and commercial vehicles are currently operating on pure or blended ethanol. Since June 1977, all service stations in Sao
Paulo have been selling a 20% ethanol-gasoline blend. In Rio de Janeiro, the
present standard mixture is 13%. It is the government's intention, however,
to increase this percentage as alcohol supplies increase.

Brazilian scientists claim there have been no problems with the vehicles
running on these blends. Their tests have shown that engines operating on
ethanol are more efficient in terms of horsepower per unit of engine weight,
and that fewer pollutants in the form of carbon monoxide and lead are
emitted.

The government has stated that if subsequent experience reveals problems,
all engines will be altered to adapt to pure ethanol. Engine conversions of
this nature are already being undertaken on a small scale. An air-cooled
straight ethanol-fueled engine has been developed and is presently being
tested as the result of a cooperative research program between Volkswagen
do Brasil and Volkswagen Research in Germany. General Motors do Brasil
is testing more than 100 ethanol-burning trucks in Brazil. Chrysler do Brasil
has also developed, and is testing, its own engines for pure ethanol operation.

Many Brazilian manufacturers recognize that ethanol has a variety of
industrial applications. It is used in the production of solvents, cosmetics,
printing inks and plastics. In cooperation with American companies, the
Brazilian government and private industry are examining different ways to
use agriculture ethanol in the production of petrochemicals for domestic and
export markets.

There is an additional incentive for the Brazilian government to promote
large-scale production of homegrown fuel other than the high cost of fuel
imports. During the past two decades, there has been a large migration of
people to the big cities, causing enormous social and economic problems. As
such, the government would like to encourage people to remain in the rural
areas. The development of an indigenous energy industry is one way to help
achieve this goal. Estimates by the government indicate that when the policy
is fully developed the energy industry will employ some 200,000 people.
This technology can be used by the local people without importing foreign
expertise and, as such, has tremendous potential.

The Brazilian government is also pursuing plans to cultivate other suitable
energy crops in addition to sugarcane. One of the most promising is the
sugar-bearing tuber, manioc. Manioc has the advantage that it grows well on
marginal land without the use of fertilizers. Brazil has much of this kind of
land (called cerrados), which includes high plains of poor agricultural quality
and generally alkaline soil. In the past, this land has been considered useless

for agricultural purposes but did serve for cattle grazing. Brazil hopes that much of the cerrados can be put to product use with the cultivation of manioc for energy purposes.

## U.S. INTERESTS IN ETHANOL ENERGY USE

There is also increasing interest in the U.S. in the use of ethanol as a gasoline additive. At present, all industrial-use ethanol is made from ethylene derived from petroleum; however, it can just as readily be made from renewable biomass materials. Renewable biomass materials include algae, agricultural crops and residues, and wood. These materials can be converted into ethanol by hydrolysis, fermentation and distillation.

Ethanol as a fuel is suitable for use in automobile engines, in power plants, for process heat generation, and as an energy source in fuel cells. Low emissions are associated with its use and, as such, it is considered appropriate for use by electric utilities as a turbine fuel for peak load requirements. In mixtures of 15-20% with gasoline, it can be used in automobile engines with essentially no carburetion modifications necessary.

The U.S. Department of Energy (DOE) has recognized the importance of an immediate, intensive program to determine ethanol's suitability as an additive to gasoline on a local and/or national basis in the near future. According to DOE, there do not appear to be any major technological barriers to either the production or distribution of alcohol fuels. The major uncertainties are availability of supply, production costs, most appropriate applications and the proper approach to commercialization. To resolve these issues, DOE is conducting a series of special investigations to examine key economic, environmental and institutional issues surrounding alcohol fuel from both coal and biomass sources.

As part of its large support of alcohol research projects, the Alternative Fuels Utilization Program of the Division of Transportation Energy Conservation is funding a variety of programs on ethanol fuels at various universities and private companies across the country.

Responsibility for most of the DOE biomass program comes under the direction of the Fuels From Biomass Systems Branch (FFB). The Branch includes in its overall biomass research several programs devoted to the investigation of ethanol fuel development.

## ETHANOL AS AN AUTOMOTIVE FUEL

Table 6.1 compares some of the properties of ethanol to those of iso-octane.

Table 6.1  Comparison Between the Properties of Isooctane and Ethanol

|  | Isooctane ($C_8H_{18}$) | Ethanol ($C_2H_5OH$) |
|---|---|---|
| Molecular Weight | 114.224 | 46.07 |
| Carbon:Hydrogen Weight Ratio | 5.25 | 4.0 |
| Carbon, % by weight | 84.0 | 52.0 |
| Hydrogen, % by weight | 16.0 | 13.0 |
| Oxygen, % by weight | 0.0 | 35.0 |
| Boiling Point, °C at 1 atm | 99.24 | 78.3 |
| Freezing Point, °C at 1 atm | -107.4 | -114.1 |
| Vapor Pressure, psia at 37.8 °C | 1.708 | 2.5 |
| Specific Gravity, 60 °F/60 °F and 1 atm | 5.795 | 6.63 |
| Coefficient of Expansion 1/F at 60 °F and 1 atm | 0.00065 | 0.00048 |
| Surface Tension, dynes/cm at 20 °C and 1 atm | 18.77 | 23 |
| Viscosity, centipoises at 20 °C and 1 atm | 0.503 | 1.17 |
| Specific Heat of Liquid, Btu/lb-F at 77 °F and 1 atm | 0.5 | 0.6 |
| Heat of Vaporization, Btu/lb at boiling point and 1 atm | 116.9 | 361 |
| Heat of Vaporization, Btu/lb at 25 °C and 1 atm | 132 | 395 |
| Heat of Combustion, Btu/lb at 25 °C Higher heating value | 20,556 | 12,780 |
| Lower heating value Liquid fuel-gaseous $H_2O$ | 19,065 | 11,550 |
| Stoichiometric Mixture, lb/air/lb | 15.13 | 9.0 |
| Autoignition Temperature, °C | 417.8 | 362.8 |
| Octane Number (research) | 100 | 106 |

The concept of employing ethanol as a fuel for the internal combustion engine is not revolutionary.  In Europe during World War II, largely as a war effort through government sponsorship, ethanol-gasoline mixes for automotive fuel were common.  The Brazilian program discussed earlier uses this method partly as a price support for their sugar crops.  The government buys up the excess sugar year by year, ferments and distills these excesses to ethanol, and blends this ethanol with the gasoline sold to the public.  For bumper crop years, as much as 16.9% ethanol has been prepared for automotive fuel.

The South African Coal, Oil and Gas Corporation Limited (SASOL) manufactures alcohol-hydrocarbon gasoline blends.[68]  Ethanol is the main constituent of their so-called motor alcohol.  SASOL has built substantial data bank information on ethanol production with over 20 years of experience. Much of this information is proprietary because the hydrocarbons blended with the motor alcohol are products of the Fisher-Tropsch process and are chemically different from petroleum-derived gasoline hydrocarbons. Other alcohols in gasoline have been investigated on a laboratory scale, and much research is still being done on alcohol-hydrocarbon blends.

There are a number of countries in addition to Brazil and the United States that are seriously investigating the economic prospect of using ethanol as a fuel source. In 1963, India sponsored a symposium on the use of ethanol in that country.[69]  In 1970, the Canadian Wheat Board sponsored a large study on the subject.[70]

Central American countries are also sponsoring various studies into the local wide-scale use of ethanol as a fuel source. In the United States, the study of wide-scale production and use of ethanol as an automotive fuel has had recurring interest. A considerable amount of work has come from the Northern Regional Research Laboratories[71-73] of the Department of Agriculture, at Peoria, Illinois. Briefly, the results have indicated that the use of ethanol in gasoline is technologically feasible but economically not competitive with petroleum products. Should the production of ethanol from agricultural and municipal wastes become efficient, economic factors may change.

Ethanol has a Btu content significantly higher than that of methanol (approximately 12,780 Btu/lb vs 9,500 Btu/lb for methanol). However, ethanol's Btu value is still significantly lower than gasoline's. A gallon of ethanol contains about 0.7 the Btu capacity of gasoline. The addition of ethanol to gasoline causes the Btu capacity to drop. As in the case of methanol, the addition of ethanol to gasoline causes an automatic leaning of the fuel mixtures because of the difference in stoichiometry of the two fuels. In addition, there is much concern and controversy as to the mpg efficiency between ethanol-gasoline blends and gasoline.

Ethanol also has a relatively high octane rating—106-107.5 RON (Research Octane Number) and 89-100 MON (Motor Octane Number). The addition of ethanol to nonleaded gasoline causes the octane rating to increase along with the antiknock capacity of the fuel.

Problems associated with the use of ethanol-gasoline blends appear to parallel those of methanol-gasoline blends. Some of these include starting, performance problems and phase separation of ethanol and gasoline because of water contamination. Problems with vapor lock and fuel system corrosion may also occur, although no actual tests have been reported in the literature.

Nebraska, with the assistance of the University of Nebraska, has recently begun examining the ethanol-gasoline blend possibilities. The initial work and results of the Nebraska Agricultural Products Industrial Utilization Committee and the gasohol program (10% ethanol in gasoline) are discussed by Scheller.[74] Scheller notes at least three factors that are important to the economic future of gasohol, namely the price of gasoline, the price of grain and the value of by-products from the alcohol manufacturing process. The first major task that the Nebraska Agricultural Products Industrial Utilization Committee performed was an extensive literature search on the use of alcohol as a fuel additive. At the time of this search, no data could be found on the use of alcohol-gasoline blends and/or exhaust gas compositions from alcohol blends in current or projected engine types. Since no extensive fleet testing had been done, no statistical data were available for a large sample of vehicle types using alcohol-gasoline blends. It was found that the economics for the fermentation process used in preparing ethyl alcohol from grain is highly sensitive to the market for the principal by-product—distillers' dried grains.[74]

Because of this, the Nebraska Agricultural Products Industrial Utilization Committee initiated several programs, including: (1) an exhaust gas analysis program to obtain precise measurement of emissions such as $CO_2$, $CO$, $NO_x$ and unburned hydrocarbons. Information on the difference in composition between exhaust gases from engines fueled with gasohol and from engines fueled with conventional gasoline are of prime interest; (2) a preliminary road test program to identify potential problems in using gasohol fuel and to provide preliminary data for the design of an extensive fleet testing program for gasohol; (3) a design for a two-million-mile fleet testing program to permit gathering statistically significant data on fuel consumption, engine and exhaust system wear, and vehicle performance with gasohol relative to conventional gasoline; and (4) a preliminary study on upgrading the value of distillers' dried grains.[74]

As in the case of methanol-gasoline blends, the problem areas for ethanol-gasoline fuel blends are relatively well-defined. Unfortunately, tests have not been adequate to offer definitive conclusions as to the extent of these problems.

## LEGAL IMPLICATIONS IN THE USE OF ETHANOL AS A FUEL (FEDERAL)

Because ethanol's primary use is as a beverage, all aspects of its production, control and use come under the control of the Bureau of Alcohol, Tobacco and Firearms (BATF) under the Department of the Treasury (as described in the appropriate sections of the U.S. Internal Revenue Code and the Code of the Federal Regulations). In light of any major development of the use of ethanol as a fuel, these laws and regulations form a significant factor that may restrict production and certainly will involve a cost.

All ethanol in the United States is closely monitored by the Bureau of Alcohol, Tobacco and Firearms to protect the $10.50/100 proof gallon tax revenue, which is derived from its production as an alcohol for human consumption. All persons having contact with ethanol, *i.e.,* distillers, warehousers, dealers, bottlers, rectifiers and users, must have prior approval from BATF for their production, handling and/or use. All production and shipment of this alcohol are recorded and closely monitored by BATF officials. To protect the interests of the government of the tax revenues due on ethanol production, individuals having contact with this alcohol usually must submit substantial bonds to the BATF before they can operate. The amounts of these bonds all but prohibit small commercial production of ethanol.

The U.S. Code does state, however, that ethanol to be used for fuel, light and power is tax free and may be removed from bonded premises after it has been denatured. Denaturization is the process of degrading ethanol and making it unfit as a beverage or for internal human medical use. There are two major classifications of denatured ethanol: (1) completely denatured alcohol, and (2) specially denatured alcohol.

Completely denatured alcohol may be produced according to two federally approved formatulations (Table 6.2). Completely denatured alcohol may be sold and used for any lawful purpose, including use as a fuel. Specially denatured alcohol can be prepared from a variety of formulations. It is a less severely denatured alcohol that is intended for specific industrial use. Consequently, it cannot be sold to the public. Table 6.3 lists those formulations that are approved for fuel use, along with a listing of product/processes using specially denatured alcohol.

In cases in which an industry desires to market ethanol to the public for use as a fuel, the processes and formulas that would be used in its production must be submitted to BATF for approval. The exact maximal proportions of ethanol that would be acceptable by the BATF as a fuel have not yet been established. Each process and formula must be analyzed by the Chemical Branch of the Bureau of Alcohol, Tobacco and Firearms. Their criteria for approval of formulas are that: (1) the formulated substance not be potable, and (2) the ethanol from the formulated substance be extracted easily for the purpose of human consumption.

The following federal regulations apply to alcohol storage, shipping and use:

Methyl alcohol is a product that requires special labeling under the Federal Hazardous Substance Act, as amended. Special labeling is required as follows:

> Because blindness and death can result from the ingestion of methyl
> alcohol, the label for this substance and for mixtures containing 4 percent
> or more by weight of this substance shall include the signal word 'danger',

the additional word 'poison' and the skull and crossbones symbol. The statement of hazard shall include 'vapor harmful' and 'May be fatal or cause blindness if swallowed'. The label shall also bear the statement 'Cannot be made non-poisonous'.[75]

### Table 6.2  Approved Formulations for Completely Denatured Alcohols

(Formula 18)
(CD 18)

To every 100 gallons of ethyl alcohol of not less than 160 proof, add:

(a)  2.50 gal of methyl isobutyl ketone
(b)  0.125 gal of pyronate or a compound similar thereto
(c)  0.50 gal of acetaldol
(d)  1.00 gal of either kerosene, deodorized kerosene or gasoline

(Formula 19)
(CD 19)

To every 100 gallons of ethyl alcohol of not less than 160 proof, add:

(a)  4.0 gal of methyl isobutyl ketone
(b)  1.0 gal of either kerosene, deodorized kerosene or gasoline

### Table 6.3  Approved Fuel Formulations for Specially Denatured Alcohols

(Formula No. 1)      To every 100 gallons of alcohol add 5 gallons wood alcohol.

(Formula No. 3-A)    To every 100 gallons of alcohol add 5 gallons of methyl alcohol.

(Formula No. 28-A)  To every 100 gallons of alcohol add 1 gallon of gasoline.

#### (Listing of Products and Processes Using Specially Denatured Alcohol)

| Product or Process | Code Number | Authorized Formulas |
|---|---|---|
| Fuel Uses, Miscellaneous | 630 | 1, 3-A, 28-A |
| Fuels, Airplane and Supplementary | 612 | 1, 3-A, 28-A |
| Fuels, Automobile and Supplementary | 611 | 1, 3-A, 28-A |
| Fuels, Proprietary Heating | 620 | 1, 3-A, 28-A |
| Fuels, Rocket and Jet | 613 | 1, 3-A, 28-A |

With regard to the storage and handling of fuels, quantities of 40,000 gallons or more gasoline or any volatile petroleum distillate or organic liquid having a vapor pressure of 1.5 lb/in.$^2$ absolute or greater under actual storage conditions must be stored in pressure tanks or reservoirs or must be stored in containers equipped with a floating roof or vapor recovery system or other vapor emission control device.[76]

With regard to shipping, both ethanol and methanol are classified as flammable liquids and, as such, require a specified "Flammable liquid" label on all shipping containers. In light of the flash point temperature of alcohol, large-scale transportation may be accomplished in tank cars and/or trucks. Special packing regulations for authorized small containment shipment are also stipulated in the regulations.

If ethanol as a fuel becomes prevalent, legislation might be implemented to require that fuel be taxed according to its Btu content. Presently, the law states that a 100% methanol-fueled vehicle would pay twice the federal tax per mile paid by the same gasoline-fueled vehicle.

## LEGAL IMPLICATIONS AT THE STATE LEVEL

In 1971, the Nebraska legislature passed a statute to develop and test the use of ethanol as a motor fuel.[77] In the statute, an Agricultural Products and Industrial Utilization Committee was established to administer a program to receive, blend and distribute ethanol-nonleaded gasoline blends to state agencies for use in their motor vehicles. BATF authorized Formula 28-A specially for denatured ethanol for this purpose. To date, about 8 cars involved in this test have been using a 10% ethanol-in-gasoline blend. The committee is now in the process or organizing a larger scale, 2-million-mile test on 30-60 of their state vehicles in conjunction with a petroleum company that will be contracted to blend and distribute this fuel.

Through this same legislation, special state taxation rates were authorized for 10% ethanol-gasoline blends and greater. This has reduced their tax-per-gallon from 8½¢ to 5½¢. In Nebraska, a 1/8¢/gal portion of the agricultural use gasoline tax is used to support the alcohol/gasoline project.

In 1973, Indiana instituted a program similar to the one in Nebraska.[78] Indiana's program is administered by the Lieutenant Governor, who is also the Commissioner of Agriculture. Of the total gasoline tax refunds returned to the farmer, 1/64th is used to support the alcohol/gasoline program.

Ethanol-gasoline blends may also meet difficulties in meeting some state fuel volatility regulations. A number of states have volatility specifications for fuels. These are summarized in Table 6.4.

## Table 6.4  State Gasoline Volatility Specifications

| State | IBP (°F, max) | 10 | 20 | 30 | 40 | FBP (°F, max) | % Recov. (min) | % Residue (max) | J | F | M | A | M | J | J | A | S | O | N | D |
|---|---|---|---|---|---|---|---|---|---|---|---|---|---|---|---|---|---|---|---|---|
| | | (max. temp, °F, for % evap.) | | | | | | | Reid Vapor Pressure (max, during months indicated) | | | | | | | | | | | |
| Alabama | | 167 | | | 392 | | | | 13.5 | 13.5 | 11.5 | 11.5 | | | 10.0 | | | 11.5 | 11.5 | 13.5 |
| Arizona | 120 | | | 284 | | | | | 13.5 | 11.5 | 11.5 | 10.0 | 10.0 | | 9.0 | 9.0 | 9.0 | 10.0 | 11.5 | 13.5 |
| Arkansas | | 176 | 221 | 284 | 392 | 437 | 95 | | | | | | | | | | | | | |
| California | | | | | | | | | | | | | | | | | | | | |
| Colorado | | 167 | | 284 | 392 | 437 | 95 | 2 | | | | | | | | | | | | |
| Connecticut | | | | | | | | | | | | | | | | | | | | |
| Florida | | 140 | | 240(9) | 365 | 437 | | 2.0 | | | | | | | 11.5 | | | | | |
| Georgia | | 149 | | 245 | 374 | 437 | 95(1) | 2 | | | | | | | | | | | | |
| Hawaii | | | | | | | | | | | | | | | | | | | | |
| Illinois | | 167 | | 284 | 392 | | | 2.0 | 15 | 15 | 13.5 | 11.5 | 11.5 | | 14.0(3) | | | 11.5 | 13.5 | 15 |
| Indiana | | 167 | | 284 | 392 | | 95 | 2 | 15 | 15 | 13.5 | 11.5 | 11.5 | | 10.0 | | 11.5 | 12 | 15 | 15 |
| Iowa | | 167 | | 284 | 392 | 437 | 95 | 2 | | | 12 | | | | 10.0 | | 12 | | | |
| Kansas | | 167 | | 284 | 392 | 437 | | 2 | | | | | | | 13.5 | | | | | |
| Louisiana | | | | 284 | 392 | 437 | | 4 | | | | | | | | | | 13.5 | | |
| Maine | | | | | | 437 | | | | | | | | | | | | | | |
| Maryland | | | | | | 437 | | 2.0 | 14.5 | 14.5 | 13.5 | 13.5 | | | 11.5 | | | 13.5 | 13.5 | 14.5 |
| Massachusetts | | | | | | | | | | | | | | | | | | | | |
| Minnesota | | 167 | | 284 | 392 | 437 | 95(4) | 2 | 13.5 | 13.5 | 13 | | | | 10 | | | 13 | | 13.5 |
| Mississippi | | 167(5) | | 284 | 392 | 437 | | 2 | 13.5 | 13.5 | 13.5(3) | | | | 9.5(3) | | | | 13.5(3) | |
| Missouri | | 167 | | 284 | 392 | 437 | | | 15.0 | 15.0 | 15.0 | 11.5 | 11.5 | 10.0 | 10.0 | 11.5 | 11.5 | 11.5 | 15.0 | 15.0 |
| Montana | | 167 | | | 392 | 437 | | | | | | | | | | | | | | |
| Nebraska | | 167 | | | 392 | 437 | 95 | 2.0 | | | | | | | | | | | | |
| Nevada | | 167 | | | 392 | 437 | | | | | | | | | | | | | | |
| New Mexico | 131 | | 221 | 284 | 392 | 464 | 95 | 2 | | | | | | | (7) | | | | | |
| New Jersey | | | | | | | 95 | | | | | | | | | | | | | |
| New York | | | | | | 437 | | | | | | | | | | | | 11.5 | 11.5 | 13.5 |
| North Carolina | | 158 | | 284 | 392 | 437 | | 2.0 | 13.5 | 13.5 | 11.5 | 11.5 | | | 10.0 | | | | | |
| North Dakota | | Max150 Min100 | | Max250 Min165 | Max375 Min265 | Max437 Min360 | 94.0 | 2.0 | | | | | | | | | | | | |
| Oklahoma | | | | | | | | | | | | | | | | | | | | |
| Rhode Island | | | | | | | | | | | | | | | | | | | | |
| South Carolina | 122 | 158 | 221 | 284 | 392 | 435 | 95 | 2 | | | 11.5 | | | | | | | 11.5 | | |
| South Dakota | | 140/149/158(2) | | 284 | 392 | | | 2 | | | | | | | 10 | | | | | |
| Tennessee | | 167 | | 284 | 392 | 437 | 95 | 2 | | 15 | 11.5 | | | | 12.5(3) | | | | | 15 |
| Texas | | 167 | | 284 | 392 | 437 | | 2 | | | | | | | | | | | | |
| Utah | | 167(6) | 267(6) | 392(6) | 392 | 437 | | 2.0 | 13.5 | 13.5 | 13.5 | 13.0 | 13.0 | | 9.5 | | | 13.5 | 14.0 | 15.0 |
| Virginia | | 158 | | 284 | | 437 | | 2.0 | 15.0 | 15.0 | 14.0 | 13.0 | 13.0 | | 10.5 | | 13.0 | 13.0 | 14.0 | 15.0 |
| Wisconsin | 131 | 158 | | | | | | 2.0 | 14.5 | 14.5 | 14.0 | | | | 11.5 | | 13.0 | 13.5 | 14.5 | 14.5 |
| Wyoming | | 167 | | 284 | 392 | | | 2.0 | | | 13.5 | 13.5 | | | 13 | | | 13.5 | 14.5 | 14.5 |

## 7

## DEVELOPMENT OF A NATIONWIDE BIOMASS-BASED ALCOHOL-GASOLINE FUEL SYSTEM

### DEVELOPMENT OF A NATIONWIDE METHANOL-GASOLINE BLEND

Projections of fuel requirements by the year 1980 have been made by various investigators. Although a 100% methanol fuel for transportation purposes is questionable, certainly a 5% methanol-gasoline blend is within short-term expectations. By the year 1990, gasoline demand is estimated to be 115 billion gallons, or 14.3 quads.[79] If alcohol-gasoline blends are used, a larger volume of fuel would be needed to meet this energy demand, as the energy content of methanol and ethanol is less than that of gasoline. The difference in volumetric demand between 5% methanol-gasoline blends and gasoline are illustrated in Figure 7.1 (this assumes that the total energy demand is constant at 14.3 Q).

As shown in Figure 7.1, a total of 5.9 billion gallons of methanol would be needed to produce 117.0 billion gallons of a 5% methanol-gasoline blend demand. This represents nearly 5 times the current industrial production of methanol from natural gas. A large methanol production plant utilizing wood for feedstock would have a daily consumption of about 1700 odt of biomass. This much capacity is comparable to a large pulp mill. Park et al.[79] notes that is an efficiency factor of 0.9 for a methanol plant of this size was assumed, the annual production would be about $65.4 \times 10^6$ gal/yr. To satisfy the projected 1990 automotive fuel demand, about 90 such plants would be needed. To supply a 10% methanol-gasoline blend, twice as many plants would have to be constructed. Bliss et al.[80] have estimated that a single 1700 odt/d plant would cost about $63.9 million. The total capital required then for 90 such plants would be $5.7 billion.

Assuming wood to be the primary biomass source used, about 49.7 million odt/yr would be needed. This translates to about 0.85 Q/yr of energy.

Methanol that would be generated from 90 plants (each having a capacity of 1700 odt/d) would produce about 0.4 Q of energy suitable for fuel use. The methanol production process would be expected to have a conversion

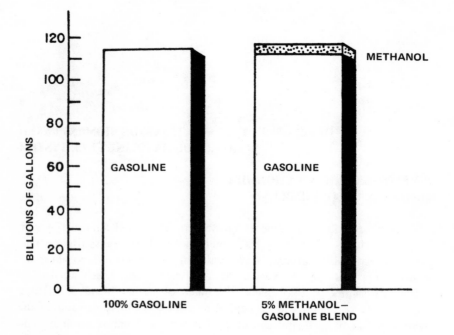

Figure 7.1 The methanol-gasoline fuel volumes required to meet 1990 energy
demands.[79]

efficiency of approximately 45%, *i.e.,* the ratio of output to input energy.
Figure 7.2 summarizes the production requirements needed for a nationwide
5% methanol-gasoline system.  To establish a system such as this by the year
1990 a tremendous growth must take place in the methanol market.

Park *et al.*[79] conclude that the effective annual growth needed would be
about 35% for the five-year period between 1985 and 1990.  The projected
growth rate is shown in Figure 7.3.   It should be noted that biomass-
generated methanol could also be supplemented by methanol derived from
coal during this growth period.  Park *et al.* conclude that a 5% methanol in
gasoline blend for the entire nation is the maximum attainable market pene-
tration by 1990.  This conclusion assumes, however, that methanol is derived
only from biomass resources.

## BIOMASS RESOURCES FOR A NATIONWIDE
## METHANOL-GASOLINE PROGRAM

Although there are several biomass sources available from which methanol
could be derived, wood residues are considered the most significant, both

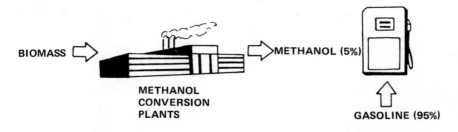

BIOMASS

METHANOL (5%)

METHANOL
CONVERSION
PLANTS

GASOLINE (95%)

Figure 7.2 The production requirements for a nationwide gasohol system based on 5% methanol-gasoline blend.

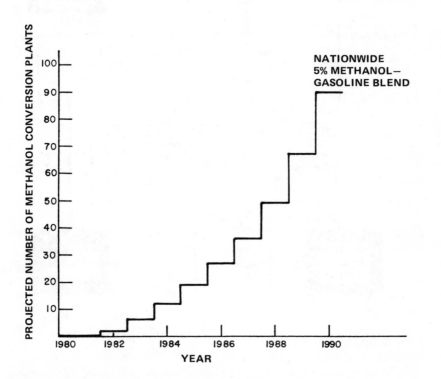

Figure 7.3 Projected growth pattern required to meet a 1990 deadline of nationwide 5% methanol-gasoline system.[79]

from a traditional point of view and in terms of quantity. Disposition of annual growth experienced in 1970 and salvaged wood from commercial forests is given in Figure 7.4. The figure shows that of the 7.0 Q of annual growth, roughly 3.4 Q were harvested in the form of logs and chips, 1.7 Q were left unharvested, and the remainder (1.9 Q) left in the forest as collection residues.

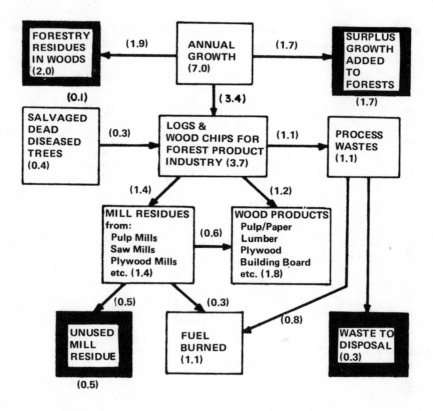

( ) – QUADS

Figure 7.4   The disposition of U.S. forest biomass based on 1970 figures. Dark portion. represent wood resources that could have been applied to energy production in 1970.[79]

A total of 3.7 Q of wood was obtained for use in the paper industry (3.4 Q harvested and about 0.3 Q of wood salvaged from diseased trees (Figure 7.4). Of this amount, 0.5 Q appeared as unused mill residues and 0.3 Q ended up as process wastes. If a 5% methanol-gasoline blend system were to be developed, 0.85 Q of wood per year would be needed. Part of the unused mill residues could be available for methanol production; however, this total resource is small. The majority of this source is likely to be used in the production of wood products and process heat/steam. The bulk of the feedstock would have to be collected from forestry residues or harvested from standing forest biomass.

As will be shown in Chapter 9, there appears to be sufficient wood available to supply 0.85 Q of feedstock in 1990. Surplus annual growth should be expected to supply less than 50% of the total requirement. In addition to this, a portion of the noncommercial standing timber on commercial forest land (which amounts to roughly 1 billion odt) is another near-term source of feedstock. Unfortunately, this reservoir is not considered an annually renewable resource.

## DEVELOPMENT OF A NATIONWIDE ETHANOL-GASOLINE PROGRAM

Figure 7.5 illustrates that roughly 5.8 billion gallons of ethanol would have to be manufactured to generate 116.8 billion gallons of gasohol to meet a 5% ethanol-gasoline blend in 1990. This figure represents approximately 29 times the current national production of industrial ethanol. A total of 86 ethanol plants capable of producing 67.3 million gal/yr would be required to produce enough ethanol for a 5% fuel blend (this assumes that an ethanol plant consumes about 12,000 ton/day of raw sugar juice and generates 207,600 gpd of ethanol). Lipinsky et al.[66] have estimated the capital cost of this size ethanol plant to be $126.8 million. Eighty-six plants would cost roughly $10.9 billion.

There are a variety of feedstock candidates besides sugarcane, including corn, wheat, sorghum, molasses, etc. Capital costs for plants based on other types of feedstock are comparable to the sugarcane case. The production requirements for a nationwide 5% ethanol-gasoline blend system are illustrated in Figure 7.6.

The primary feedstock candidate for ethanol production is agricultural waste and residues. Estimates show that there are roughly 100 million acres of unused cropland currently available in the United States.[81] It is not likely, however, that all of this land could be put to effective food production for alcohol manufacture.

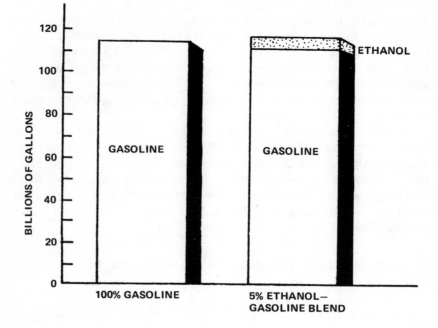

Figure 7.5  The ethanol-gasoline fuel volumes required to meet 1990 energy demands.[79]

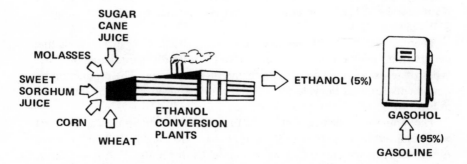

Figure 7.6  The production requirements for a nationwide gasohol system based on 5% ethanol-gasoline blend.

Lipinsky *et al.*[66] have estimated that the maximum potential sugarcane production in the continental U.S. could be on the order of 26 million dry tons annually (dt/yr) by 1990. On a volume basis, this corresponds to about 1.6 billion gal/yr of anhydrous ethanol (roughly 28% of the amount required to obtain a 5% ethanol-gasoline blend in the country). This means that only 25 of the necessary 86 ethanol plants could use sugarcane juice as feedstock. In addition, the continental U.S. can only supply one sugarcane crop per year. As such, the storage of feedstocks for ethanol production would be a major consideration and perhaps a limiting factor in the feasibility of this approach. Another possible feedstock is grain. Grains have the advantage in that long-term crop storage does not present a serious problem. Lipinsky *et al.* have proposed that in a critical need situation, 20 million acres could be added to our national corn production capacity. A bushel of corn can be converted to approximately 2.8 gallons of ethanol and, at 83 bu/ac (average for the U.S., 1973-1975), the expected annual yield from 20 million acres would be 4.6 billion gallons of ethanol. This is about 80% of the amount needed for a nationwide 5% ethanol-gasoline blend.

Production of sweet sorghum, sugar beets, wheat and other grains could also be increased as feedstock for ethanol production. Sufficient additional feedstock could be produced from sugar crops, corn and other grains to make the ethanol required for a 5% alcohol-gasoline blend, nationally. Figure 7.7 summarizes potential feedstock contributions for ethanol production.

## THE ECONOMICS OF ALCOHOL PRODUCTION

Market prices in 1975 of various alcohols are given in Table 7.1. The equivalent cost of these alcohols per $10^6$ Btu are given in Table 7.2. Alcohols are relatively high in cost in comparison to gasoline. As such, the additional costs of an alcohol-gasoline system would result in higher consumer prices for alcohol-gasoline blends.

By way of review, methanol can be made from several different processes using a variety of feedstocks. These are summarized in Table 7.3. Various estimates of methanol production from coal[82-86] indicate capital costs of $32,500 (±$1,430)/ton/day of methanol production capacity and a methanol cost of 11.7 (±0.3) ¢/gal based on a coal price of 30¢/million Btu and total annual capital charges of 15% of investment.

A rough estimate of the cost of methanol from coal (in ¢/gal), based on 1975 economics, can be obtained from the following equation:

$$Cost = 11.7 \ (0.00977 \ A + 0.222 \ B) \qquad (7.1)$$

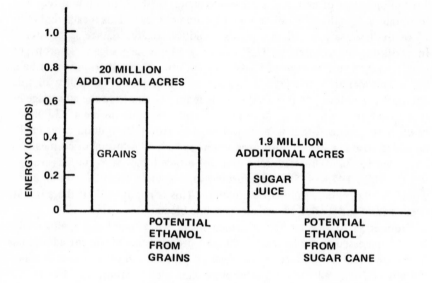

Figure 7.7 Potential available resources for ethanol production.

Table 7.1 Market Prices (1975) for Various Alcohols[82]

| Alcohol | Current Market Price ($/gal) |
|---|---|
| Methanol | 0.38-0.49 |
| Ethanol | 1.04-1.14 |
| *n*-Propyl Alcohol | 1.57 |
| Isopropyl Alcohol | 0.63-0.70 |
| *n*-Butyl Alcohol | 1.15-1.49 |
| Isobutyl Alcohol | 1.07-1.47 |
| *sec*-Butyl Alcohol | 1.43 |
| *tert*-Butyl Alcohol | 1.75 |

Table 7.2 Equivalent Alcohol Market Costs per $10^6$ Btu of Energy[82]

| Alcohol | Current Market ($/million Btu) Price |
|---------|--------------------------------------|
| Methanol | 5.99-7.72 |
| Ethanol | 12.19-13.47 |
| n-Propyl Alcohol | 15.98 |
| Isopropyl Alcohol | 6.93-7.70 |
| n-Butyl Alcohol | 10.95-14.18 |
| Isobutyl Alcohol | 10.32-14.19 |
| sec-Butyl Alcohol | 13.72 |
| tert-Butyl Alcohol | 17.34 |

where A is the cost of coal in ¢/million Btu, and B is the total annual capital charges, i.e., percentage of total investment. Similarly, methanol cost from gas is given by[87]:

$$Cost = 12.6(0.00905 N + 0.00673 B + 0.850) \qquad (7.2)$$

where N is the natural gas cost (¢/million Btu).

Ethanol can be prepared either by fermentation (for human consumption) or by synthesis (for industrial use). In general, cost estimates for methanol production from petroleum and coal are well-documented. Estimates for ethanol production by enzymatic and acid hydrolysis methods are, however, only preliminary and have not been verified.

## GOVERNMENT PARTICIPATION IN ESTABLISHING A NATIONWIDE GASOHOL FUEL SYSTEM

To ensure that an alcohol-gasoline system would be economically viable, prices of gasoline and alcohol-gasoline blends must be competitive. When two fuel types are equivalent in all performance aspects, their prices are expected to be the same. The cost differences of alcohol-gasoline blends and gasoline are small on a cents per gallon basis; however, on a national scale, these differences are greatly magnified.

The annual expense of converting from nonleaded gasoline to 5% ethanol-gasoline blend was estimated by Park et al.[79] to be $6.9 billion. The imported oil saved by the conversion to the blended fuel, in terms of energy, would be 0.49 Q.

Three other cases were considered in Park's economic analysis:

Table 7.3  Cost for Methanol Production

| Method | Feedstock | Feedstock Cost | Plant Capacity | Total Capital Cost | Capital Cost per Unit Capacity | Annual Operating Cost | Methanol Cost (¢/gal) | Methanol Cost $/10^6 Btu August 1974 | Methanol Cost $/10^6 Btu November 1975 | Investigator |
|---|---|---|---|---|---|---|---|---|---|---|
| Solid Waste Pyrolysis Synthesis | Municipal trash | -$5/ton $0/ton +$5/ton | 400 ton/day methanol | $77.6 x 10^6 | $194,000/ton/day $10,100/10^6 Btu/day | -$0.21 x 10^6 +$2.54 x 10^6 +$8.04 x 10^6 | 18.3¢/gal 25.1¢/gal | $2.87 $3.94 $6.06 | $4.31 $5.91 $9.09 | Wiatrak,1974 |
| Koppers-Totzek ICI Synthesis | Eastern bituminous A coal | $7.50/ton | 5000 ton/day methanol | $204.5 x 10^6 | $41,000/ton/day $2140/10^6 Btu/day | $48.26 x 10^6 | 15.6¢/gal 19.5¢/gal | $2.44 $3.06 | $3.66 $4.59 | Jaffee, 1974 |
| Koppers-Totzek ICI Synthesis | Western bituminous B coal | $3/ton | 5000 ton/day methanol | $204.5 x 10^6 | $41,000/ton/day $2140/10^6 Btu/day | $38.15 x 10^6 | 13.5¢/gal 17.4¢/gal | $2.12 $2.73 | $3.18 $4.10 | Jaffee, 1974 |
| Winkler ICI Synthesis | Western sub-bituminous coal | $3/ton | 5000 ton/day methanol | $196.6 x 10^6 | $39,300/ton/day $2050/10^6 Btu/day | $35.85 x 10^6 | 12.9¢/gal 16.5¢/gal | $2.02 $2.59 | $3.03 $3.89 | Jaffee, 1974 |
| Lurgi ICI Synthesis Single Pass Methanol Conversion | Western sub-bituminous coal | $3/ton | 5000 ton/day methanol + 185x10^6 scfd SNG | $425.3 x 10^6 | --- $1570/10^6 Btu/day | $64.68 x 10^6 | 9.0¢/gal 11.9¢/gal | $1.41 $1.87 | $2.12 $2.81 | Jaffee, 1974 |
| Lurgi ICI Synthesis Methane Separation before Methanol Synthesis | Western sub-bituminous coal | $3/ton | 5000 ton/day methanol + 80 x 10^6 scfd SNG | $289.3x10^6 | --- $1685/10^6 Btu/day | $50.63 x 10^6 | 10.4¢/gal 13.5¢/gal | $1.63 $2.12 | $2.45 $3.18 | Jaffee, 1974 |
| Hydrocarbon Oxidation | Natural gas | 40¢/10^6 | 5000 ton/day methanol | $43.15 x 10^6 | $8630/ton/day $452/10^6 Btu/day | $30.90 x 10^6 | 6.9¢/gal | $1.09 | $1.64 | U.S. Energy Outlook, 1973 |
| Steel Production By-product | Carbon monoxide from coal | Cost at $8/ton | 22,080 ton/day | $485 x 10^6 | $22,000/ton/day $1150/10^6 Btu/day | $160 x 10^6 | 11.9¢/gal | $1.87 | $2.81 | Steinman, 1974 |

1. a nationwide 5% methanol-gasoline blend;
2. a 10% methanol-gasoline blend applied to one-half of the national gasoline demand in 1990; and
3. a 10% ethanol-gasoline blend applied to one-half the national gasoline demand in 1990.

Figure 7.8 summarizes the estimated costs obtained from the study. Note that the costs reported are the total program costs divided by the estimated number of barrels of imported oil saved in each case (in 1976 dollars).

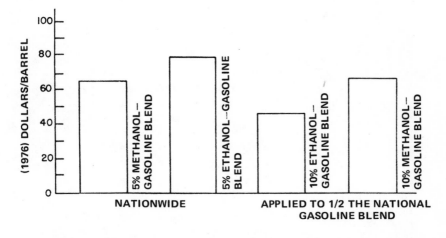

Figure 7.8 The magnitude of the nationwide costs of various alcohol-gasoline programs by 1990.[79]

From this study and others, it should be evident that the cost of transportation and processing systems required to develop and maintain a national alcohol-gasoline fuel system is significant.

With the small percentages of alcohol in the alcohol-gasoline blends, there would have to be a significant change in the price of the alcohols to produce even small changes in the price of blends. Government incentives, such as investment tax credits for biomass alcohol production, would help, although the magnitude of impact this would have in reducing the price of an alcohol-gasoline blend is not clear.

## 8

# SPECIAL USES AND PROBLEMS WITH ALCOHOL FUELS

## FUEL USES OF PROPYL AND BUTYL ALCOHOLS

Propyl and butyl alcohol have been used as fuels on a limited basis. *N*-propyl alcohol has been employed as an antifungucide in jet fuel. Isopropyl alcohol has been used in high-compression aircraft engines as an antidetonation injection fluid with water and also as a deicer for automobile fuel.

*Tert*-butyl alcohol has had more direct use in automobile fuel than the others. Atlantic Richfield Company has used *tert*-butyl alcohol concentrate as a blending component for gasoline under the trade name ARCONOL®. Atlantic Richfield[88] notes that the blend concentration used is at a nominal 5% level, which insures freedom from haze formation and depreciation in vehicle drivability. For ARCONOL concentrations above 5%, haze formation occurs if blending takes place with ARCONOL-free fuels. Commercial use of ARCONOL at the 5% level has resulted in no haze problems. Richfield claims, from vehicle drivability testing with 5% blends in a wide variety of base gasolines, that no depreciation in drivability occurs. In some fuels, improved drivability was found. At higher concentrations (10 and 15%), drivability depreciation has been observed in some vehicles.

Octane blending performance has been documented in a number of leaded and clear gasoline and was found to be favorable. Carburetor antiicing performance, when compared to other known antiicing materials, showed ARCONOL to be very effective.

ARCONOL blends with gasoline in a nonideal volume relationship, which results in a net increase in fuel volume. This moderately increases the Reid Vapor Pressure, the front end volatility (distillation range) and the vapor: liquid ratio of the fuel. However, Richfield claims that the increased volatility does not result in increased vapor locking tendencies.

Tests have been made to examine the effect of ARCONOL-containing gasolines on the metals and elastomers present in automotive fuel systems. It has been noted that ARCONOL in gasoline can increase fuel corrosiveness to zinc and magnesium in the absence of effective nonphosphorous corrosion inhibitors. However, there are several commercially available inhibitors that can be used. Although ARCONOL fuels induce marginal swelling and softening increases in elastomers used in fuel systems, results were within specification allowances. It should be noted that the use of 5% ARCONOL decreases the Btu/gal value of gasoline by about 1%. However, test data on loss of power and fuel economy were found to be no worse than this loss in heating value.

The Goodyear Tire and Rubber Company disclosed in 1974 a patent on a *tert*-butyl alcohol-gasoline-water blend. The blend is reported to have much better cold temperature properties than ethanol- or methanol-gasoline blends.

A major advantage of *tert*-butyl alcohol over methanol and ethanol is that *tert*-butyl alcohol-gasoline mixtures can accept much more water in solution before phase separation. This alcohol is, however, much more unattractive from an economic standpoint for large-scale production.

## ALCOHOLS FOR ELECTRIC POWER GENERATION

Recently, there has been growing interest in the use of methanol as a fuel for electric power generation. The first major test was run by the Vulcan-Cincinnati Corporation (VCC) in New Orleans in 1972.[88] The combustion demonstration test was held at the A. B. Paterson Station of New Orleans Public Service (NOPSI) in September of 1972, sponsored by NOPSI, Southern California Edison Company, Consolidated Edison Company (New York) and Vulcan-Cincinnati Corporation. The test was monitored by 24 American, Japanese and European companies, representing utility, engineering, construction and trading interests.

The test facility consisted of a commercial utility boiler with a capacity of 50 MW. The test showed that methanol provides not only a good stable flame and good burning efficiency, but also better performance than natural gas or fuel oil in reducing pollutant nitrogen oxides. In this test, methanol was burned satisfactorily in existing utility boilers originally designed for alternate firing with or without fuel oil. This is significant in that no modifications to these boilers would be required.[89]

Several advantages and observations were noted from the use of methanol as a fuel for electric power generation[88]:

1. No particulates were observed from the discharge stack.
2. Nitrogen oxides found in the fuel gases were less than those detected from natural gas combustion and significantly less than those detected from oil combustion.
3. The carbon monoxide concentration from methanol firing was less than that observed from oil and gas firing.

4. No sulfur compounds were emitted during methanol combustion.
5. Analyses for aldehydes, acids and unburned hydrocarbons indicated
   that only trace quantities were emitted during methanol combustion.

In addition to these, soot deposits in the furnace from previous oil firing were burned off by methanol combustion. This promoted higher heat-transfer rates during the combustion process.[88]

The General Electric Company performed a series of tests using methanol as a fuel for their MS7000 gas turbine.[90] For a base case fuel, number two distillate was used, against which all comparisons were made. Three types of methanol fuel were studied: 100% dry methanol, an 80% methanol and 20% water mixture, and a 75% distillate and 25% methanol blend. To compensate for the Btu difference in all these fuels, the actual mass flows of the fuel into the turbine were varied accordingly. For example, in the case of pure methanol, approximately twice as much fuel was pumped into the turbine than for the number two distillate fuel. Due to these differences, some mechanical advantage was gained in power output with the increased mass flow of the methanol fuels. The emission results obtained from the study are illustrated in Figure 8.1.

In terms of regulatory levels, the CO maximum emissions as proposed by the U.S. Environmental Protection Agency (EPA) are 900% greater than those of number two distillate and, as such, the CO emissions are well within this level.

General Electric reports that some retrofitting is required before turbine generators can operate on methanol fuel. Problems arise from the low lubricity of methanol. Pumps built to transfer fuel oil use the fuel itself for their own lubrication. Effective lubrication would not be achieved with methanol, and pump retrofitting would be necessary.

If methanol is prepared in a clean form, it is believed that few corrosion difficulties will be associated with its use as a fuel for electric power generation. On the other hand, methanol derived from coal will have sulfur contamination, which can cause corrosion and air pollution problems. Similarly, if the methanol used is shipped from overseas and proves to be contaminated with seawater, the sodium that would be present in the fuel could cause major corrosion difficulties.

In addition to the concept of methanol as a fuel for fixed electric power generators, interest has been shown in the use of methanol as a fuel for fuel cells. These fuel cells fall into two categories: (1) those that use the methanol directly as the fuel in the electrolyte, and (2) those that use the methanol as a source of hydrogen. In the latter case, methanol is converted to hydrogen via steam reforming.

Military fuel cells in this country predominantly use hydrazine as fuel. The disadvantage with hydrazine is that it is highly toxic. As such, should fuel cells become available to the public, methanol would be much more desirable than hydrazine as a fuel.

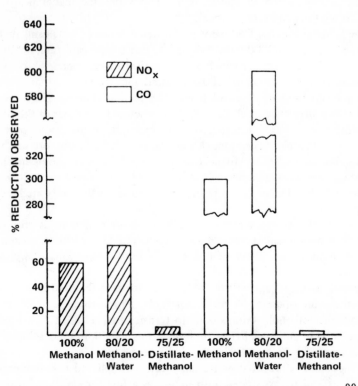

Figure 8.1 Emission results obtained from the General Electric Co. tests.[90]

## THE USE OF METHANOL IN THE NATURAL GAS INDUSTRY

Estimates show that roughly 16% of the annual U.S. consumption of natural gas matches the amount of energy wasted each year from the natural gas that is flared to the atmosphere in the Middle East.[89] The potential use of this energy has been under consideration for some time. Probably the largest problems associated with the use of this energy are in transportation and storage. No pipelines exist to transfer the natural gas from the Middle East sources and, as such, this gas must be converted to some alternative form before it can be transported economically to the consumer.

Natural gas can be liquefied [Liquefied Natural Gas (LNG)] by the application of pressure and very low temperatures. As LNG, it can then be shipped in liquid form to the U.S., as is done presently. To reduce the high costs involved in LNG and tanker facilities, conversion of natural gas to methanol is being considered. Methanol can be shipped to the U.S. in ordinary tankers and then reconverted to methane for use in the natural gas system. McGhee[91] has noted several advantages to this approach:

1. Methanol can be shipped in conventional tankers that are used for hauling crude oil or refined products. LNG, on the other hand, requires much more expensive, specially designed, single-purpose ships.
2. Methanol can be shipped to any port and offloaded into conventional terminal tankage used for crude oil, refined products, petrochemicals, etc.
3. Methanol can be delivered from such a terminal through existing distribution facilities, *e.g.*, pipelines, rail cars or tank trucks.
4. Methanol can be stored inexpensively for long periods of time for peak load or seasonal demand uses, *e.g.*, home heating and agriculture operations.

Disadvantages to this approach are that methanol synthesis plants are capital cost-expensive and would be outside of U.S. direct control. Further, methanol to methane conversion plants would add substantial costs and be energy-intensive. The direct use of the methanol fuel would be more practical. The economic feasibility of methane-methanol-methane process also depends on the removal of the import duty of this fuel-directed methanol. The details of this current legislation are given by Baratz *et al.*[61]

Another potential use of methanol for the natural gas industry might be through the use of coal as a feedstock. Baratz *et al.*[61] suggest that a combination plant producing methanol and synthetic natural gas from coal could possibly serve a dual purpose. A plant of this type could provide some gas needs for the industry, while the excess for methanol production could be stored for conversion to methane and used in the natural gas system in high-demand periods.

## MISCELLANEOUS USES OF ALCOHOL FUELS

Research into the use of alcohols in diesel engines has been revived. Early studies[92,93] showed that increased power and reduced smoke emission result when alcohol is sprayed into the intake air. Alcohols were not effectively mixed in the fuel itself.

Another potential use suggested for alcohols is related to the transportation and processing of coal slurries. The extraction of coal requires substantial amounts of water, which may be scarce in the Western coal region. The transportation of water-coal slurries via pipeline to energy-demand areas where water supplies are plentiful has attracted recent interest. If the coal is transported in a water slurry, the coal drying process at the pipeline ends can be highly energy-intensive. Methanol has been suggested as a fluid for carrying the coal in these pipelines.[94] The coal would not require drying at its destination. Also, the methanol would enhance the Btu content of the mixture.

Alcohols may have future use as replacements for natural gas in drying processes. This might also include the significant amounts of energy consumed in grain-drying operations.

## A SUMMARY OF ENVIRONMENTAL PROBLEMS

Various environmental advantages and disadvantages of the use of alcohol as a fuel have been discussed throughout the book. Some general comments about the environment and alcohol fuel are appropriate at this point.

Methanol as a fuel for the automobile engine appears to have environmental advantages; namely, $NO_x$ emissions can be made significantly lower. For methanol-gasoline blends, the environmental effects are not fully understood. Ethanol-gasoline blends for automobile fuel do not appear to show any improvement in reduction of emissions over straight gasoline. By contrast, the results from electric power generation seem promising. No sulfur problems and very reduced $NO_x$ emissions with the use of 100% methanol as a fuel were observed in combustion tests. However, as with every new concept, tradeoffs must be considered with increased alcohol fuel use.

Aldehyde emissions have been observed to increase when using alcohols in automobiles. The effects of wide-scale aldehyde dispersions are not known yet and require in-depth study.

The toxicology of alcohols under controlled conditions is reasonably well-documented. Unfortunately, under uncontrolled situations, as would be the case if alcohols had a wide-scale use as fuel, the effects to the populace are not known. Potential vapor problems may arise with the alcohols in unventilated areas, and certainly fire hazards are another major consideration.

One major unknown is the effect on the environment caused by spillage of alcohols. Very little information exists on the effects on marine life, should there be a large-scale spillage from ocean-going tankers. Since alcohols are completely soluble in water, cleanup of a spill of alcohol in the ocean or other waterway may be virtually impossible. A summary of these and other potential environmental problems is given in Figure 8.2.

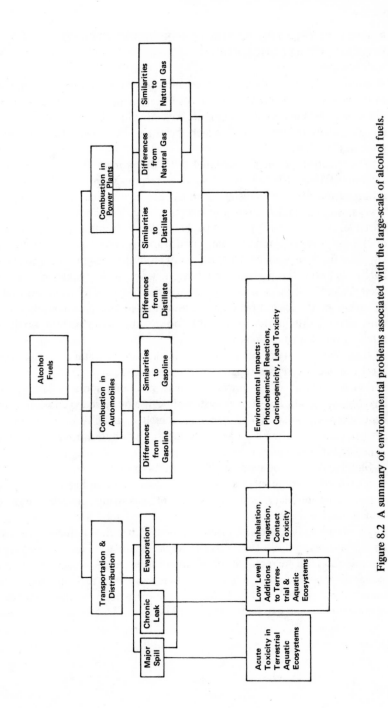

Figure 8.2  A summary of environmental problems associated with the large-scale of alcohol fuels.

## PROBLEMS RELATED TO THE TRANSPORTATION AND STORAGE OF ALCOHOL FUELS

Both ethanol and methanol have vapor pressures and densities comparable to those of gasoline. As such, the transportation and storage of these alcohols are much the same as with gasoline, *e.g.*, on ships, in rail tank cars, in tank trucks and in storage tanks. Alcohol can also be transported through steel pipelines.

The greatest problem associated with the storage and transportation of alcohol destined for use as a fuel is water consumption. Alcohol has an affinity for water, and substantial quantities of water are present in gasoline vessels, storage tanks, ships and barges. Since gasoline and water are not miscible, this does not present a problem. Water forms a layer at the bottom of the gasoline tank and is usually undisturbed. This layer of water is actually beneficial, as it catches any sediment from the gasoline and also is a good barrier between the gasoline and bottom leaks in the tanks. If a leak does occur in the bottom of an oil or gasoline tank, the water leaks out first, and discovery of the leak is possible before actual loss of product. By contrast, with alcohol this barrier would not form. Any water present would enter into solution with the alcohol.

If alcohol is blended with gasoline, the amount of water in a tank becomes a critical factor. Phase separation problems associated with water contamination are severe. Water prevention systems would add a major expense to any transportation and storage system. Because of these expenses, if a nationwide alcohol-gasoline blend becomes a reality, alcohol and gasoline might have to be transported separately through a distribution network and finally mixed at the fuel pump.

Should alcohols be used separately from gasoline, then small amounts of water would not pose serious problems. If, on the other hand, contamination is from seawater, increased corrosion problems would probably result from alcohol used as a fuel.

As noted earlier, alcohol-gasoline blends experience reduced Btu content in comparison to hydrocarbon fuels. As an example, twice the quantity on a volume basis will be required for methanol to perform the same requirements as gasoline or fuel oil. This means that twice the fuel quantity must be shipped and stored in the transportation network. Various studies that address these issues can be found in the literature.[95,96]

## 9

## MASS PRODUCTION OF BIOMASS FOR SYNTHETIC FUELS

## OVERVIEW OF BIOMASS SPECIES AND PRODUCTIVITY

Photosynthesis is the process by which green plants utilize solar energy to fix carbon dioxide for transformation into organic material. This process has afforded man a source of food, fiber and fuel throughout his history. Man's ability to manipulate this resource to his particular advantage has provided the major impetus to his evolution as a dominant species. Today, faced with the predicament of ever-dwindling supplies of fossil fuels, mankind must reexamine this resource in the hope of extending its utility still further. He must fashion from this renewable, carbonaceous feedstock, forms of energy that can be effectively and economically substituted for petroleum and natural gas and their derivative products.

A brief discussion of the various types of biomass was given in Chapter 1. By way of review, the biomass sources that could be used include:

- manures collected from confined livestock operations,
- accumulated crop residues from primary processing operations,
- discarded crop residues left in the fields after harvesting,
- wood and bark residues, which accumulate from primary wood manufacturing operations,
- wood and bark residues from logging operations, and
- unused standing forests having no mixed commercial value.

### Agricultural and Feedlot Wastes

Anaerobic fermentation of wastes has been practiced for more than 100 years; however, little advance in the understanding of the application to agricultural wastes has occurred since about 1935. It is estimated that this clean renewable source of energy could supply a large fraction of the energy used in farming operations.[97] In addition, marginal lands might be used to produce biomass for fuels.

Jewell et al.[97] have estimated the total energy use on 40 and 100 cow dairies and 1000-head feedlots to be $164 \times 10^6$, $307 \times 10^6$ and $670 \times 10^6$

kcal/yr, respectively (this excludes energy used in the manufacture of equipment and farm chemicals). The estimated maximum annual methane energy that could be generated in these operations was estimated as $216 \times 10^6$, $473 \times 10^6$ and $2280 \times 10^6$ kcal, respectively. Jewell *et al.*'s estimates show that a dairy farm could produce more energy than it consumes, and a feedlot could produce more than three times the quantity consumed.

The generation of methane gas from organics has been studied extensively for more than 100 years, and many full-scale operating systems were installed on farms in Europe and other areas in the 1930s and 1940s. Unfortunately, none were installed in this country, nor was adequate information available to estimate the possibility of using them.

By 1935, pilot-scale agricultural waste fermentors and full-scale sewage sludge anaerobic digestors were in use. Subsequently, a number of different types of reactors were constructed, mainly on farms throughout Europe. These units combined various designs, and systems were operated under a variety of conditions. Parameters such as mixing, heating and substrate concentration varied from unit to unit. Unfortunately, few recorded data are available to provide an analysis of the impact of these modifications, which emphasizes the need for further studies to optimize this technology.

Figure 9.1 illustrates the distribution of feedlot wastes. Jewell *et al.*[97] note that the huge amounts of wastes generated in beef feedlots of as many as 100,000 cattle indicate that these would be areas favored for consideration of this technology. It should be noted, however, that there are only a few facilities of this size. Roughly 83% of all beef installations have fewer than 1000 head. There are several hundred thousand dairies in the U.S., with many of them having between 40 and 100 cows. This means that for anaerobic fermentation to be widely used as a means of syngas production, it must be feasible for the small dairies and feedlots with fewer than 1000 head.

The potential benefits from using anaerobic fermentation of agricultural wastes include: energy production, labor reduction, pollution control, potential volume reduction, aesthetic value, residue recovery for refeeding and other uses, and nutrient conservation. Factors that detract from these benefits include costs for: generation, gas cleaning and storage, energy conversion, and the quantity of wastes not utilizable. Because of site-specific parameters, generalizations cannot be readily made.

In general, manures exist in relatively small quantities. Similarly, collected agricultural residues represent only a moderate overall energy contribution. There are constraints that restrict the energy potential of agricultural residues. Crop residues that are left in the field have a value to farmers both as: (a) a source of plant nutrients, and (b) as a soil conditioner that improves soil tilth and water-holding capacity. In addition, crop residues have a low density per unit land area, which implies high biomass collection and transportation costs.

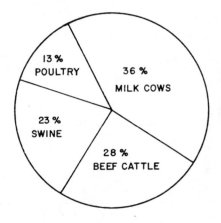

Figure 9.1 The breakdown of feedlot wastes.[97]

## Logging Residues

Typically, logging residues are distributed over remote regions, which also suggests high collection and transportation costs. Logging residues are largely within the realm of the forest industry, which means that perhaps 50% or more of this material would be used in the manufacture of high-value wood products on collection. This same constraint also applies to the use of standing commercial forests.

## Overall Conditions

The annual domestic energy consumption is approximately 78 Q of energy feedstock.   The breakdown of domestic energy resources is given in Figure 9.2. Roughly 36 Q of petroleum are utilized annually. Of this amount, approximately 50% is removed from proven domestic reserves of 200 Q, *i.e.*, 31.7 trillion bbl of oil. Annual natural gas consumption amounts to about 23 Q, and almost all of this is removed from proven domestic reserves. Coal reserves (amounting to 4100 Q) contribute about 14 Q to annual consumption.   Hydroelectric sources provide about 3.1 Q and nuclear power 1.5 Q, on an annual basis. Estimated undiscovered reserves of fossil fuels, natural gas and fissionable fuels far exceed proven reserves in each case; however, considerable uncertainty prevails in these estimates.

There is roughly 570 Q contained in U.S. standing forests, which consist of about 55% of commercial timber and 45% of noncommercial timber. The

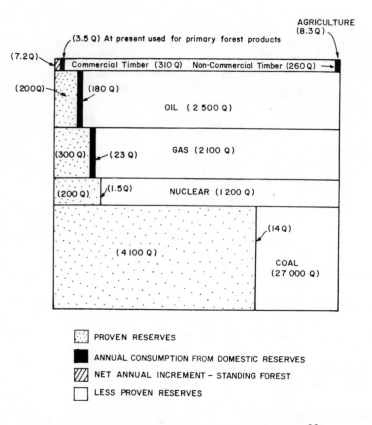

Figure 9.2  The breakdown of domestic energy resources.[98]

annual growth of all domestic forests is roughly 10.7 Q, and about one-third of this is utilized for the manufacture of primary products.[98]  Agriculture generates 8.3 Q each year, which is equally divided between primary products, *i.e.,* food, feed and fiber and crop residues, primarily from corn and small grains.  The breakdown of agricultural biomass is given in Figure 9.3. About half the annual growth of existing forests can be considered as a potential energy sources.  This amounts to about 2.5 Q of noncommercial timber in commercial forests, 1.5 Q in noncommercial forests, 0.7 Q in logging residues and 0.5 Q in unused mill residues.  The remaining portion (5.5 Q) of the annual increment is either used in commercial products or has commercial use potential and, as such, is not considered as potential energy feedstock.  This quantity consists of harvested and nonharvested commercial timber, mill residues used for primary product or as fuel, and potentially usable logging residues. The energy potential from existing crops is 2.2 Q, which consist of two-thirds the unused crop residues generated each year. The combined energy potential from existing forest and agricultural crops amounts to about 7.4 Q/yr.[98]

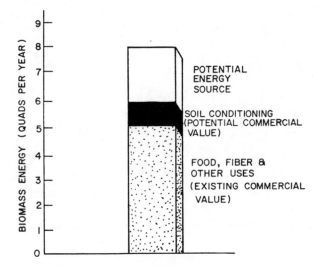

Figure 9.3 The breakdown of agriculture-biomass resource in terms of annual quad-equivalents.

The regional distribution of existing biomass sources depends on the type of source. Figure 9.4 gives a regional distribution of agriculture residues generated annually. Most of this is corn and small grain stover, occurring in the central corridor of the grain-producing states. The regional distribution of logging residues is shown in Figure 9.5(A), and standing forest distribution is given in 9.5(B). At present, the rate of removal of commercial timber in the northwest exceeds the total annual growth of that region.

## LAND SUITABILITY, AVAILABILITY AND SITE CHARACTERISTICS

Theoretically, it is possible to cultivate biomass almost anywhere, if sufficient resources are expended. For practical purposes, however, careful consideration of land sites will be needed for large-scale biomass production to be used as energy crops. Salo *et al.*[99] have noted the following criteria for land suitability for biomass production:

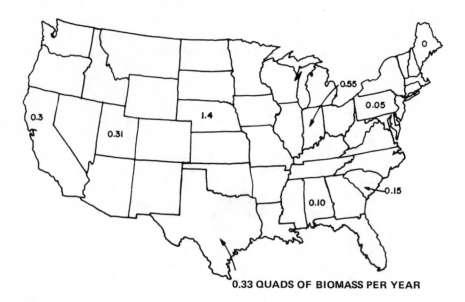

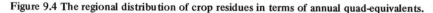

0.33 QUADS OF BIOMASS PER YEAR

**Figure 9.4 The regional distribution of crop residues in terms of annual quad-equivalents.**

1. a minimum of 25 in. of annual precipitation,
2. arable land, and
3. slope equal to, or less than, 30% (17°).

Precipitation, land quality and terrain have a direct impact on crop yield, management practices and costs. These are broad characteristics and their use provides only quantitative approximations of suitable land. Annual precipitation is only one aspect of moisture balance relationships. For example, distribution during the year and evapotranspiration considerations also have a direct effect.

Precipitation is an important indication of a region's capacity to support high levels of productivity. Odum[100] approximated the climatic biotic communities that can be expected with different, evenly distributed amounts of annual moisture in temperate latitudes (Table 9.1).

Annual rainfall distributions across the country are illustrated in Figure 9.6. It should be noted that total rainfall alone does not determine the biotic situation. Other factors, such as its annual distribution, potential evapotranspiration, and the amount of moisture resulting from fog, must also be considered. The impact of these other factors can be observed in different locations across the country. For example, the average annual rainfall in Florida is 53 inches, but irrigation of cropland is also required. Figure 9.7

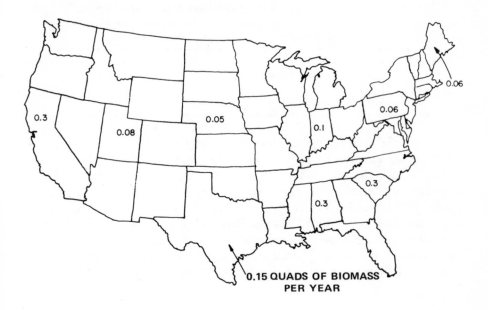

0.15 QUADS OF BIOMASS
PER YEAR

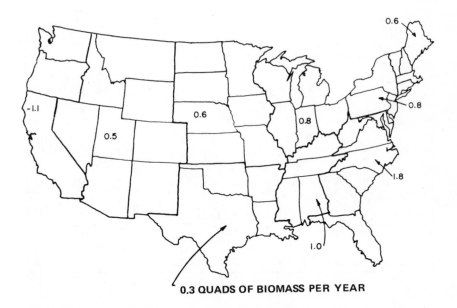

0.3 QUADS OF BIOMASS PER YEAR

Figure 9.5  The regional distribution of (A) logging residues and (B) standing forests, in quad-equivalents.

Table 9.1. Rainfall Distributions that can be Expected for Various Regions.[100]

| Land Description | Rainfall (in./yr) |
|---|---|
| Desert | 0-10 |
| Grassland or Open Woodland | 10-30 |
| Dry Forest | 20-50 |
| Wet Forest | >50 |

illustrates the average annual potential evapotranspiration across the U.S.

There are a number of other climatic factors that are also important in site selection. Regions with mild climates are not only desirable because they support good growth, but also because they would afford much easier mechanical harvesting and transportation of green biomass chips or bales. Potential sites in Northern New England and the Northern Great Lakes Region are less attractive than those further south because of short growing seasons and severe winters, which reduce annual productivity. These factors could result in management and logistical problems.

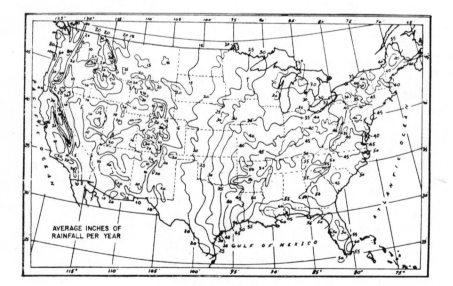

Figure 9.6 The annual distribution of precipitation in the continental U.S.[101]

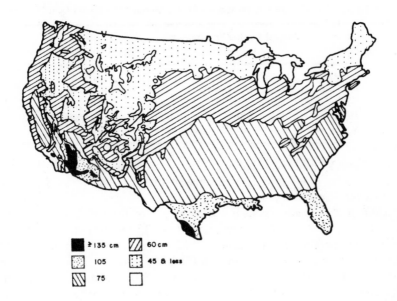

Figure 9.7 The average annual potential evapotranspiration in the continental U.S.

High isolation regions with warm, dry climates, such as the southwestern states, are also unlikely to sustain high yields of many fast-growing biomass species because of arid conditions. These regions are also limited in irrigation potential. Pressure on the southwest's water resources will continue to grow as the demands of industry, agriculture, population and emerging energy technologies increase. This will also reduce water availability and increase its cost. On the other hand, if problems associated with supplying sufficient water to the area at a reasonable price are overcome, the potential of the southwest as a source of energy crops should be reevaluated. A few drought-hardy genera, such as the Eucalyptus tree, could well be used in this region as a biomass crop, if water becomes more available.

Several classifications have been developed to rate land suitability. The U.S. Forest Service (USFS) has used annual productivity to differentiate between commercial and noncommercial forest land. The USFS classification also subdivides the commercial land into five separate categories. In this system, timber production capacity is the only criterion used in site classification. The categories are summarized in Table 9.2.

Table 9.2.  U.S. Forest Service Site Classes:  Commercial Forest Land[102]

| Site Class | Annual Timber Productivity ($ft^3$/ac-yr) | Approximate dte/ac-yr |
|---|---|---|
| I | 165 or more | 2.8 or more |
| II | 120-165 | 2-2.8 |
| III | 84-120 | 1.4-2.0 |
| IV | 50-85 | 0.8-1.4 |
| V | 20-50 | 0.3-0.8 |

For Class V, land incapable of producing more than 50 $ft^3$/ac-yr, species respond poorly to intensive forest management. For Classes I and II, the land produces at least 120 $ft^3$/ac-yr.

Another classification system was developed by the Soil Conservation Service (SCS) to encourage the proper use of land. This system indicates the suitability of soils for various purposes. Eight broad groupings or capability classes have been designated. Soils in Class I have few limitations restricting their use. Those in Class VIII cannot be used for commercial plant production and have limited value for other purposes, such as recreation and wildlife habitat. Each of these classes is described in Table 9.3.

Capability subclasses are soil groups within each class. They are designated by adding a small letter—e, w, s or c—to the class numeral, e.g., IIe. The letter e notes erosion as the principal limiting factor; w indicates that water on or in the soil interferes with plant growth or cultivation; s represents a limitation due to the shallow, droughty or stony nature of the soil; and c indicates a climatic limitation.

Arable land, or that which is suitable for cultivation, is generally included in SCS Classes I-IV. It should be noted, however, that 5% of the country's cropland is in the lower classes (V-VIII). Because the system is relatively subjective, land types included in various classes and subclasses could differ by county or state; however, it is the most comprehensive system of land categorization developed.

The availability of land suitable for producing biomass solely for the purpose of energy production, or at least in part, will depend on competing uses of biomass species and future cropland forest land trends. At present, there are about 385 million acres of cropland. Cropland is being abandoned at a rate of 2.7 million acres each year.[104] This is occurring primarily in the southern and eastern regions of the country and is a result of land obsolescence, partly caused by low soil fertility. Cropland is also being transferred to urban, transportation, reservoir and various other nonagricultural uses.

At the same time, new cropland is being formed at an annual rate of roughly 1.3 million acres through such practices as irrigation, drainage, land clearing and improved harvesting, irrigation and fertilization techniques.

Table 9.3  Land Capability Classes—Soil Conservation Service[103]

| Class | Capability and Precautions | Primary |
|-------|---------------------------|---------|
| **(Suitable for Cultivation)** | | |
| I | Excellent land, flat, well-drained. Suited agriculture with no special precautions other than good farming practice. | Agriculture |
| II | Good land with minor limitations, such as slight slope, sandy soils or poor drainage. Suited to agriculture with precautions such as contour farming, strip cropping, drainage, etc. | Agriculture Pasture |
| III | Moderately good land with important limitations caused by soil, slope or drainage. Requires long rotation with soil-building crops, contouring or terracing, strip cropping or drainage, etc. | Agriculture Pasture Watershed |
| IV | Fair land with severe limitations caused by soil, slope or drainage. Suited only to occasional or limited cultivation. | Pasture Tree crops Agriculture Urban-industrial |
| **(Not Suitable for Cultivation)** | | |
| V | Land suited to forestry or grazing without special precautions other than normal good management. | Forestry Range Watershed |
| VI | Suited to forestry or grazing with minor limitations caused by danger from erosion, shallow soils, etc. Requires careful management. | Forestry Range Watershed Urban-Industrial |
| VII | Suited to grazing or forestry with major limitations caused by slope, low rainfall, soil, etc. Use must be limited and extreme care must be taken. | Watershed Recreation Wildlife Forestry Range Urban-industrial |
| VIII | Unsuited to grazing or forestry because of absence of soil, steep slopes, extreme dryness or wetness | Recreation Wildlife Watershed Urban-industrial |

The SCS[105] has indicated that nearly 110 million acres of noncropland have high-to-medium potential as cropland during the next 10-20 years. The distribution of this potential land is illustrated in Figure 9.8.

Figure 9.8  Noncropland regions having high-to-medium potential for conversion to cropland.

Commercial forest land consists of about 485 million acres.[106]  An additional 15 million acres of forest land is in the form of parks, wilderness and various other uses not compatible with commercial forestry.  Spurr and Vaux[106] have projected a reduction in total U.S. forest lands to 455 million acres by the year 2020.  Even with this loss, they suggest that forest productivity will increase substantially.

Projections by Cotner[104] and Spurr and Vaux[106] indicate that fewer than one million acres of cropland and forest land will be converted to urban, transportation or reservoir uses annually.  Remaining agricultural and forest land would be available for other uses, and much of it could be used to increase agricultural and forestry production.  Reliable projections of the extent to which world demands for food and fiber must be met by agriculture and forestry in the U.S. are difficult to make.  The land available for biomass production in the future will largely depend on the balance between food and fiber demands, as well as national energy needs.

Actual land use patterns are established by economic, legal, institutional and political considerations. It does appear, however, that sizable amounts of suitable land could be available for biomass production, if current land use trends and projected productivities continue.

## THE ENERGY FARM CONCEPT

Although a vast source of biomass feedstock is potentially available for synfuel production, a systematic and economical approach to harvesting this supply must be devised for a nationwide bioconversion system to be feasible. The concept of energy farms, with the purpose of biomass production of energy feedstock as a primary product, has been under investigation. The energy farm concept can be applied to both agricultural and forest sources. Application of energy farming for procuring silvicultural biomass for energy conversion is only discussed here; however, the basic principles for both agriculture and silviculture are very similar.

The silvicultural energy farm has been envisioned by several authors/ investigators[98,99,107] as consisting of plantings of selected, rapidly growing tree species, at relatively close spacings. Crop harvesting would take place at appropriate intervals or rotations, and succeeding crops would be produced by coppicing, *i.e.,* sprouting from stumps.

This concept does put a restriction on the specific biomass species that should be grown. For example, most conifers do not coppice and, as such, plantings would have to be restricted to selected hardwood species. Note that the advantage of coppicing is that it precludes the need to replant after each harvest. The energy farm would require intensive crop management, including fertilization, irrigation, and weed and pest control. In this sense, energy crop production would parallel field crop production, rather than conventional forestry.

A number of factors will determine the extent to which energy farms can impact on the nation's energy demands. The two most important factors are yield or productivity, and land availability.

Productivity levels, *i.e.,* the yield produced per unit time-space, will vary with the species planted, the cultural practices used, and the conditions specified by the farm site, including climate, weather and soil characteristics. These factors may be controlled and/or modified to some extent. Promising species can be selected through experimentation and improved by genetic recombination, and cultural practices may also be selected and optimized through experimentation. This will provide cost-effective approaches to energy crop management. Crop management practices will vary, depending on the species and the specific site. Site-specific factors can be either modified by cultural practices, such as irrigation, fertilization and liming, or accommodated by appropriate species selection. The success in optimizing biomass yield will largely be a function of the time and research dollars spent on these issues.

Land availability will also play an important role in the success or failure of energy farm systems. With current productivity levels of short-rotation tree crops, the production of one quad of energy (in the form of biomass) requires 5-12 million acres of land.[97] Energy crop farming, therefore, will require large amounts of land to make a significant national impact on the energy supply. The acquisition of land for biomass production is restricted by current competing land uses in those regions where large biomass farms might conceivably be located. As the demands for food and fiber products increase, competing land uses may become even more restrictive, if these increased demands are not accompanied by proportional increases in food/ fiber production efficiency. Land availability is heavily dependent on socio-economic factors only tangentially related to energy farming. If the energy farm concept is to attain a respectable status, the land required for its purpose must either be obtained from lower-valued land of minor use in agriculture and forestry, or the value of energy products must increase to the point where biomass farming can compete for land with high-value uses.

Energy farms are not yet in existence; however, investigators have analyzed and evaluated this concept by estimating biomass production costs. The major areas that must be better defined to apply such a system include:

1. Species composition—the types of tree species likely to comprise the biomass crop;
2. Productivity—the annual average yield attainable per acre;
3. Production—the total quantity of biomass produced in a farm-year;
4. Farm size—a function of production and productivity, coupled with the ability to acquire and manage land;
5. Land acquisition—both the method and extent to which land may be acquired, as well as its suitability for biomass production and its cost;
6. Farm configuration—the spatial relationship between planted acreage and the surrounding land;
7. Crop management—practices that might be utilized in the optimization of productivity or to ensure a desired level of production;
8. Operations—the spectrum of activities needed in production, their sequence, logistics and seasonal limitations;
9. Operations specifications—the equipment, labor and material required for operations, and their costs;
10. Site location—the effect of site-specific conditions and characteristics on various parameters; and
11. Time frame—the effect of anticipated or potential technological developments on productivity, operations, costs, etc.

Tree species that comprise conventional forest crops are usually selected on the basis of desirable wood properties. Other criteria for selection include rapid growth and ease of establishment. Tree species for energy farms must be based on a short-rotation production scheme; that is, the major criteria for selection will be the species' ability to grow rapidly. There are a number of factors or general guidelines that will assist the silviculturist in his selection of short-rotation crops. The major ones include:

1. *Rapid Juvenile Growth* where trees are grown in an intensive mode. Selected tree species must get off to a rapid growth and avoid a lag period after field establishment.
2. *Ease of Establishment and Regeneration* including a variety of hardwood species whose stands can be established with cuttings, and regeneration can be achieved by coppicing after harvesting.
3. *Maintenance Free* so that species chosen should be free from major insect and fungal pests. Trees should be chosen for their resistance to attack from pests. Genetic diversity and susceptibility to pests can be exploited to build in resistance to attack.

In general, conifers (softwoods) do not offer as wide a selection as do hardwood species for energy crops. The softwoods do not satisfy short-rotation criteria; however, there are exceptions. The majority of conifers do not display rapid juvenile growth. Furthermore, true coppice reproduction of softwoods is primarily limited to redwoods.

Conifers should not be eliminated as energy crop candidates. Nursery and field planting techniques for softwoods are well-established. From a technical standpoint, then, stand establishment and regeneration should not pose major problems. In addition, conifers are usually less site-demanding than hardwoods, and could therefore be started on moist and dry upland sites, which have medium heavy-to-sandy soils.

As previously noted, many hardwood trees fit the short-rotation criteria well. Hardwoods generally coppice well. Many can be established by cuttings, rather than seedlings, and exhibit rapid early growth. The disadvantage with hardwoods is that they do not exhibit great site adaptability. Their full productive potential can usually be achieved only on inherently fertile, well-drained and aerated sites, with an adequate supply of moisture throughout the growing season. Intensive crop management would be needed for farms situated on less-desirable sites.

There are a number of candidate species for silvicultural biomass farms. Some of the most promising species are the following:

1. *American Sycamore,* which is found in all states east of the Great Plains except Minnesota.
2. *Eucalyptus* spp., generally found in the frost-free areas of the southeast and in California.

3. *Loblolly Pine,* which grows in the Coastal Plain and Piedmont, from Delaware and Central Maryland south to Florida and west to eastern Texas.
4. *Populus* spp., which includes:  Eastern Cottonwood, which grows in southern Quebec and Ontario to southeastern North Dakota, south to western Kansas, western Oklahoma, southern Texas, northwestern Florida and Georgia; and *Black Cottonwood,* which is found south along the Pacific Coast from Kodiak Island and southeastern Alaska to mountains in southern California.  These species also extend eastward into southwestern Alberta, south-central Montana, central Idaho, northern Utah and Nevada.
5. *Sweetgum,* which extend from Connecticut southward throughout the east to central Florida and eastern Texas.  The species is also found as far west as Missouri, Arkansas and Oklahoma, and north to southern Illinois.
6. *Tulip-poplar,* which is found throughout the eastern part of the country, from southern New England, west to Michigan, and south to central Florida and Louisiana.
7. *Red alder,* which is confined to the Pacific Coast region from southeastern Alaska, extending south through Washington, northern Idaho and western Oregon to Santa Barbara, California.

Most of these species exhibit rapid juvenile growth, which is the prime criteria for short-rotation biomass production.  General remarks about the site characteristics and requirements for these species are given in Table 9.4.

Table 9.4.  Site Requirements for Various Tree Species

| Species | Comments on Site Characteristics |
| --- | --- |
| American Sycamore | Species is tolerant to large variations in groundwater levels. Generally requires plentiful supply of groundwater. |
| Eucalyptus spp. | Only a few species are frost-resistant. |
| Loblolly Pine | Grown on a variety of soils ranging from flat, poorly drained, groundwater podsols of the Coastal Plain to the old residual soils of the upper Piedmont. |
| Eastern Cottonwood | Survives on infertile sands and relatively stiff clays. Does not grow well more than 15-20 ft above stream levels. |
| Black Cottonwood | Grows well on some upland soils, if sufficient moisture is available. |
| Sweetgum | Tolerant of different soils and sites. |
| Red Alder | Grows on soils ranging from gravel to sand and clay. Soils not considered a serious limitation, except as they affect availability of soil moisture. |

## THE ECONOMICS OF BIOMASS-ENERGY PRODUCTION

Economic analyses and projections on biomass production costs via large-scale energy farms have been performed by various investigators.[98,99] Inman[98] estimated production costs for 10 study sites. The study showed that a major factor influencing production costs is the level of productivity. They found that as productivity increased, cost decreased. It should be noted that this effect is exhibited both between sites within different levels of productivity, and by varying the productivity at given sites.

Many parameters will affect the cost of biomass production for energy. Productivity will play a dominant role. The higher the productivity, the less acreage needed to produce a given tonnage of biomass. Changes in productivity will reflect on various cost items, which are determined on the basis of acreage. These cost items or categories include irrigation, land lease, land clearing and preparation, planting and weed control. Cost categories generally independent of acreage considerations include fertilization, planning, supervision, field support, access roads, loading and harvesting. Transportation can be considered to lie between these two cost groups. While trucking costs rely to an extent on farm acreage (inasmuch as it affects the average hauling distance), the amount of biomass that must be transported is the major cost determinant. In addition, certain elements of fertilization costs are controlled by acreage. These include the costs of lime, as well as the cost of ground application of lime and fertilizer during the first year of each rotation. Production costs can also be expected to vary with the rotation period, since average annual average productivity varies with the age of the crop at the time of harvest.

The time frame or rotation period will also impact on productivity. Therefore, its selection must be based on a careful consideration of the variations in site conditions, as well as the species selected for crops. It can be expected that the optimum rotation time for one species at a given site may not be optimum for another site/species combination. As such, production costs are likely to vary considerably for different species/site selections.

## SYSTEMATIC UTILIZATION OF FORESTRY RESIDUES

Wood and bark residues are generated in relatively large quantities in mills that process timber into primary products, *i.e.*, lumber and plywood. They are also generated, along with some foliage residues, in the growing and harvesting of timber. These residues can provide a ready source of energy, especially for the forest products industry. They can also be considered a source of raw materials for a number of products, such as wood pulp and particleboard.

Mill residues have heating values that depend on the species of wood in-volved, the relative contents of wood and bark, and, most important, the moisture content. Heating values of representative North American commer-cial wood species range from about 7900-9700 Btu/lb on a moisture-free basis. The heating values of bark are slightly higher than for wood of the same species. Moisture content of mill residues may range from 2-8% for sander dust to 75% for green bark. The combustion efficiency of oven-dried wood is estimated to be about 82%. However, for moisture contents as high as 50%, the combustion efficiency is only about 68%.

Unlike most coals and heavy fuel oils, wood and bark residues have negli-gible sulfur content and, as such, combustion does not present a sulfur oxide emission problem. Furthermore, wood and bark have relatively low ash con-tents in comparison to coal. Wood and bark have high oxygen contents, which means that less oxygen has to be supplied from air for combustion.

Estimates of mill residues generated annually were given in an earlier chap-ter. Representative prices of mill residues are presented in Table 9.5. Prices of wood and bark residues tend to vary widely. They exhibit considerable variability with respect to location. Some mill residues now being left unused are available at zero or nominal cost and, in some situations, are considered to be a nuisance.

Table 9.5  Some Representative Prices Paid for Wood and Bark Mill Residues ($/million Btu)

|  | West Coast | South | Maine |
|---|---|---|---|
| Chips |  |  |  |
| Softwood |  |  |  |
| 1975 | 1.47-2.21 | — — | — — |
| 1976 | 2.35 | 1.47-1.65 | 2.35 |
| Hardwood |  |  |  |
| 1976 | — — | 1.06-1.18 | — — |
| Shavings |  |  |  |
| 1975 | 0.22-0.32 | — — | — — |
| 1976 | — — | 0.44 | — — |
| Sawdust |  |  |  |
| 1975 | 0.15 | — — | — — |
| 1976 | — — | 0.09-0.12 | 0.29 |
| Bark |  |  |  |
| 1975 | 0.14 | — — | — — |
| 1976 | 0.34 | 0.06 | — — |

Roughly 75% of all wood residues generated in 1970 by the lumber, plywood and miscellaneous wood products industries were utilized for some purpose. Approximately 56% of all wood residues generated were used for nonenergy products, primarily for wood pulp. The remainder of the used wood residues (about 13 million dte), constituting 19% of total wood residues, were either used as fuel where produced or sold as fuel. Approximately 60% of the bark residues were put to some use in 1970. Probably most was used for energy production.

Demands for mill residues are likely to increase rapidly, both the energy and nonenergy products, in the forest products industry itself. The allocation of such residues among uses will depend in the long-term on the relative prices of residue-based forest products and alternative fuels. It is unlikely, although possible, that mill residues will be available for energy uses outside the forest products industry. It should be noted, however, that cogeneration of electricity is a strong possibility. There are a number of forest products plants in the country that will occasionally sell power to the local grid.

Another potentially large source of biomass for energy production are forest residues. The term "forest residues" refers to logging residues, intermediate cuttings, understory removal and annual mortality. Annual mortality refers to trees killed by natural agents. These trees are generally widely dispersed, hence making it likely that their collection for any purpose would be uneconomical in most cases.

Intermediate cuttings involve the removal of small or inferior trees from a stand for the purpose of stand improvement. Understory removal refers to the removal of shade-tolerant shrubs and trees growing beneath the canopy of an older commercial forest. Both these operations could provide biomass for energy.

Logging residues probably represent the only widely and readily available source of forest residues. These are the leftovers of commercial logging operations and generally consist of the aboveground portions of trees not removed in such logging operations, *i.e.*, branches, foliage and pieces of stems too small or too defective for conversion into conventional wood products.

Conceivably, forest residues could be collected and reduced to fuel sizes by chipping or processing through a hammermill similar to those used for reducing mill residues to fuel. The total aboveground logging residues generated in the U.S. in 1970 were estimated to be about 83 million dte. Stump-root systems left as logging residues in 1970 were estimated at 104 million dte.[98]

On a per-acre basis, logging residues are generally most concentrated in the Pacific Northwest, where much of the timber harvested is derived from stands of large, over-mature, old-growth timber. Trees in these stands are often rotten and subject to breakage on falling, with broken or rotten pieces being left onsite.

Forest residues generally have no real value to forest landowners at present. As such, it may be assumed that they are currently available at no cost. Current use of forest residues in the U.S. is negligible.

In recent years, during periods of high pulp chip prices, there have been cases in which larger-sized residues were salvaged and chipped in the Pacific Northwest. This has generally taken place in the national forests, where timber cutting contracts require that logging residues be collected and piled. Thus, collection costs are subsidized, with the user paying only for chipping and transportation.

Another area of residue use is really a reduction in residue production. This is occasioned by whole-tree chipping, wherein entire trees, foliage, etc., are reduced to pulp chips by mobile chippers. The ability and willingness of pulp mill operators to accept chips of this kind are limited. It is an increasing practice, however, and probably will continue to be as fiber becomes more expensive.

Utilization of forest residues, particularly aboveground logging residues, is likely to increase over the next several decades, for both energy and nonenergy uses. Properly processed forest residues can be put to many of the same uses as mill residues. The admixture of wood and bark is a problem in many nonenergy uses. The separation of bark from chipped residues is, however, a subject of current research, and significant progress has been made.

If the value of fiber for wood pulp and reconstituted building materials increases, and the value of residue biomass for energy increases, manufacturers will probably turn to much of what is now being left as residues in the forests. This will manifest itself both in changes in utilization standards and in increased efforts to collect residues after primary harvesting operations.

Table 9.6 gives the U.S. Department of Agriculture's (USDA) forest statistics on potentially available wood resources on a state-by-state basis. The statistics include only the wood that would be available after the needs of the forest products industry are satisfied. Wood resources potentially available could supply needed feedstock for a 10% ethanol in gasoline fuel program. It should be noted that such a program would utilize roughly 60% of the total wood resource and, as such, the feedstock supply could be vulnerable because of competition for the resource by other potential uses. Table 9.6 indicates that a national 10% methanol in gasoline fuel program can be supplied by potentially available wood resources.

## CONCLUSIONS

A nationwide program of biomass-derived fuel production could have favorable direct and indirect consequences for the country. Some of the expected benefits that would result from a large-scale biomass-based fuel program include an improvement in trade deficit, a decrease in foreign oil dependence, and a reduction in the rate of consumption of the world and national oil reserves. The successful implementation of a nationwide program of biomass-based fuel production will, however, depend on a number of critical elements, including the availability of sufficient biomass resources to implement the program and the cost of biomass feedstock.

Table 9.6  Potential Wood Feedstock Resource by States Based on 1976 Statistics
Adopted from the U.S. Dept. of Agriculture, Forest Service, Forest Statistics of the U.S. 1977, Review Draft (1978)

| Regions States | Mill Residues 10⁶ dt/yr | Commercial Forest Land Forest Residues 10⁶ dt/yr | Commercial Forest Land Surplus Growth 10⁶ dt/yr | Commercial Forest Land Annual Mortality 10⁶ dt/yr | Non commercial Timber 10⁶ dt/yr | Non commercial Forest Land Reserved & Deferred 10⁶ dt/yr | Non commercial Forest Land Unproductive 10⁶ dt/yr | Total Wood 10⁶ dt/yr | Total Wood Q/yr | Total Methanol Equivalent[a] 10⁹ gal/yr | Total Percent of Total |
|---|---|---|---|---|---|---|---|---|---|---|---|
| **Northeast** | | | | | | | | | | | |
| Connecticut | 0.006 | 0.118 | 1.412 | 0.353 | 0.235 | 0.059 | 0 | 2.183 | 0.037 | 0.29 | 0.24 |
| Delaware | 0.013 | 0.118 | 0.176 | 0.059 | 0.059 | 0 | 0 | 0.245 | 0.007 | 0.03 | 0.05 |
| Maine | 0.388 | 3.471 | 6.706 | 6.294 | 3.294 | 0.294 | 0.118 | 20.565 | 0.350 | 2.74 | 3.92 |
| Maryland | 0.074 | 0.588 | 1.059 | 0.353 | 0.471 | 0.118 | 0 | 2.663 | 0.045 | 0.36 | 0.51 |
| Massachusetts | 0.015 | 0.235 | 2.353 | 0.412 | 0.647 | 0.118 | 0 | 3.780 | 0.064 | 0.50 | 0.72 |
| New Hampshire | 0.049 | 0.529 | 4.059 | 0.353 | 1.059 | 0.118 | 0.059 | 6.226 | 0.106 | 0.83 | 1.19 |
| New Jersey | 0.015 | 0.059 | 0.353 | 0.235 | 0.176 | 0 | 0 | 0.838 | 0.014 | 0.11 | 0.16 |
| New York | 0.215 | 1.353 | 3.176 | 2.941 | 3.118 | 1.647 | 0.118 | 12.568 | 0.214 | 1.68 | 2.40 |
| Pennsylvania | 0.280 | 2.235 | 14.000 | 2.235 | 3.235 | 0.235 | 0.059 | 22.279 | 0.379 | 2.97 | 4.25 |
| Rhode Island | 0.002 | 0 | 0.294 | 0.059 | 0.059 | 0 | 0 | 0.414 | 0.007 | 0.06 | 0.08 |
| Vermont | 0.049 | 0.529 | 0.882 | 0.706 | 1.353 | 0.059 | 0 | 3.578 | 0.061 | 0.48 | 0.68 |
| Total: 10⁶ dt/yr | 1.106 | 9.235 | 34.470 | 14.000 | 13.706 | 2.648 | 0.354 | 75.519 | — | 10.05 | 14.38 |
| Q/yr | 0.019 | 0.157 | 0.586 | 0.238 | 0.233 | 0.045 | 0.006 | — | 1.284 | — | — |
| **Northern Plains** | | | | | | | | | | | |
| Kansas | 0.030 | 0.059 | 0.176 | 0.353 | 0.471 | 0 | 0.059 | 1.148 | 0.020 | 0.15 | 0.22 |
| Nebraska | 0.016 | 0.059 | 0.176 | 0.059 | 0.118 | 0 | 0.059 | 0.487 | 0.008 | 0.06 | 0.09 |
| North Dakota | 0.004 | 0 | 0.059 | 0.235 | 0.118 | 0 | 0 | 0.416 | 0.007 | 0.05 | 0.08 |
| South Dakota | 0.069 | 0.118 | 0.529 | 0.118 | 0.118 | 0 | 0.059 | 1.011 | 0.017 | 0.13 | 0.19 |
| Total: 10⁶ dt/yr | 0.119 | 0.236 | 0.940 | 0.765 | 0.825 | 0 | 0.177 | 3.062 | — | 0.39 | 0.58 |
| Q/yr | 0.002 | 0.004 | 0.016 | 0.013 | 0.014 | 0 | 0.003 | — | 0.052 | — | — |
| **Corn Belt** | | | | | | | | | | | |
| Illinois | 0.128 | 0.588 | 0 | 0.412 | 0.059 | 0.059 | 0 | 1.246 | 0.021 | 0.17 | 0.24 |
| Indiana | 0.130 | 0.471 | 1.000 | 0.294 | 0.471 | 0.059 | 0 | 2.425 | 0.041 | 0.32 | 0.46 |
| Iowa | 0.059 | 0.353 | 0 | 0.176 | 0.353 | 0.059 | 0 | 1.000 | 0.017 | 0.13 | 0.19 |
| Missouri | 0.275 | 1.118 | 0.294 | 0.294 | 3.471 | 0.118 | 0.059 | 5.629 | 0.096 | 0.75 | 1.07 |
| Ohio | 0.208 | 0.882 | 1.941 | 0.765 | 0.824 | 0.059 | 0 | 4.679 | 0.080 | 0.62 | 0.89 |
| Total: 10⁶ dt/yr | 0.800 | 3.412 | 3.235 | 1.941 | 5.176 | 0.354 | 0.059 | 14.979 | — | 1.99 | 2.85 |
| Q/yr | 0.014 | 0.058 | 0.055 | 0.033 | 0.088 | 0.006 | 0.001 | — | 0.255 | — | — |
| **Southeast** | | | | | | | | | | | |
| Alabama | 0.786 | 5.176 | 10.235 | 2.471 | 2.706 | 0.059 | 0 | 21.433 | 0.364 | 2.86 | 4.09 |
| Florida | 0.171 | 2.118 | 5.353 | 1.647 | 2.000 | 0.118 | 0.353 | 11.760 | 0.166 | 1.57 | 2.24 |
| Georgia | 0.503 | 7.765 | 15.882 | 3.765 | 3.294 | 0.706 | 0 | 31.915 | 0.543 | 4.26 | 6.09 |
| South Carolina | 0.360 | 3.529 | 5.588 | 1.765 | 2.824 | 0.118 | 0 | 14.184 | 0.241 | 1.89 | 2.71 |
| Total: 10⁶ dt/yr | 1.820 | 18.588 | 37.058 | 9.647 | 10.824 | 1.001 | 0.353 | 79.291 | — | 10.58 | 15.13 |
| Q/yr | 0.031 | 0.316 | 0.630 | 0.164 | 0.184 | 0.017 | 0.006 | — | 1.348 | — | — |
| Subtotal: 10⁶ dt/yr | 3.845 | 31.471 | 75.703 | 26.353 | 30.531 | 4.003 | 0.943 | 172.849 | — | 23.01 | 32.94 |
| Q/yr | 0.066 | 0.535 | 1.287 | 0.448 | 0.519 | 0.068 | 0.016 | — | 2.939 | — | — |

| | | | | | | | | | | | |
|---|---|---|---|---|---|---|---|---|---|---|---|
| **Appalachian** | | | | | | | | | | | |
| Kentucky | 0.246 | 1.176 | 5.765 | 0.941 | 1.353 | 0.176 | 0 | 9.657 | 0.164 | 1.29 | 1.84 |
| North Carolina | 0.435 | 4.647 | 11.118 | 3.176 | 3.412 | 0.647 | 0 | 23.435 | 0.398 | 3.12 | 4.47 |
| Tennessee | 0.143 | 1.647 | 9.059 | 1.118 | 2.294 | 0.412 | 0 | 14.673 | 0.249 | 1.96 | 2.80 |
| Virginia | 0.362 | 3.471 | 7.235 | 2.647 | 3.941 | 0.529 | 0 | 18.185 | 0.309 | 2.42 | 3.47 |
| West Virginia | 0.175 | 1.647 | 7.118 | 1.471 | 2.412 | 0.176 | 0 | 12.999 | 0.221 | 1.73 | 2.48 |
| Total: $10^6$ dt/yr | 1.361 | 12.588 | 40.294 | 9.353 | 13.412 | 1.941 | 0 | 78.949 | — | 10.52 | 15.06 |
| Q/yr | 0.023 | 0.214 | 0.685 | 0.159 | 0.228 | 0.33 | 0 | 1.341 | — | — | — |
| **Southern Plains** | | | | | | | | | | | |
| Oklahoma | 0.081 | 0.588 | 0.765 | 0.235 | 0.824 | 0 | 0.941 | 3.434 | 0.058 | 0.46 | 0.66 |
| Texas | 0.503 | 3.059 | 5.706 | 1.176 | 2.235 | 0.059 | 2.412 | 15.150 | 0.258 | 2.02 | 2.89 |
| Total: $10^6$ dt/yr | 0.584 | 3.647 | 6.471 | 1.412 | 3.059 | 0.059 | 3.353 | 18.584 | 0.316 | 2.48 | 3.55 |
| Q/yr | 0.010 | 0.062 | 0.110 | 0.024 | 0.052 | 0.001 | 0.057 | — | — | — | — |
| **Delta States** | | | | | | | | | | | |
| Arkansas | 0.576 | 4.588 | 7.000 | 1.706 | 3.118 | 0.059 | 0 | 17.047 | 0.290 | 2.27 | 3.25 |
| Louisiana | 0.630 | 3.941 | 8.235 | 2.529 | 2.941 | 0.059 | 0 | 18.335 | 0.312 | 2.44 | 3.50 |
| Mississippi | 0.680 | 4.706 | 7.941 | 1.941 | 2.647 | 0.059 | 0 | 17.974 | 0.306 | 3.20 | 3.43 |
| Total: $10^6$ dt/yr | 1.886 | 13.235 | 23.176 | 6.176 | 8.706 | 0.177 | 0 | 53.356 | 0.908 | 7.91 | 10.18 |
| Q/yr | 0.032 | 0.225 | 0.394 | 0.105 | 0.148 | 0.003 | 0 | — | — | — | — |
| **Lake States** | | | | | | | | | | | |
| Michigan | 0.190 | 1.647 | 9.941 | 6.765 | 1.706 | 0.353 | 0.059 | 20.661 | 0.351 | 2.75 | 3.94 |
| Minnesota | 0.131 | 1.059 | 6.412 | 3.000 | 0.824 | 0.588 | 0.353 | 12.367 | 0.210 | 1.65 | 2.36 |
| Wisconsin | 0.198 | 2.176 | 3.471 | 1.471 | 1.412 | 0.059 | 0.059 | 8.846 | 0.150 | 1.18 | 1.69 |
| Total: $10^6$ dt/yr | 0.519 | 4.882 | 19.824 | 11.235 | 3.942 | 1.000 | 0.471 | 41.874 | 0.711 | 5.58 | 7.99 |
| Q/yr | 0.009 | 0.083 | 0.337 | 0.191 | 0.067 | 0.017 | 0.008 | — | — | — | — |
| **Pacific** | | | | | | | | | | | |
| California | 1.347 | 7.353 | 0 | 3.176 | 2.588 | 2.059 | 5.000 | 21.523 | 0.366 | 2.87 | 4.11 |
| Oregon | 0.737 | 13.824 | 0 | 9.471 | 5.824 | 1.471 | 1.059 | 32.386 | 0.551 | 4.32 | 6.18 |
| Washington | 0.539 | 9.059 | 0 | 7.647 | 5.000 | 4.176 | 0.706 | 27.127 | 0.461 | 3.62 | 5.18 |
| Total: $10^6$ dt/yr | 2.623 | 30.236 | 0 | 20.294 | 13.412 | 7.706 | 6.765 | 81.036 | 1.378 | 10.81 | 15.47 |
| Q/yr | 0.045 | 0.514 | 0 | 0.345 | 0.228 | 0.131 | 0.115 | — | — | — | — |
| **Mountain** | | | | | | | | | | | |
| Idaho | 1.232 | 2.882 | 6.235 | 2.706 | 3.176 | 3.706 | 1.235 | 21.172 | 0.360 | 2.82 | 4.04 |
| Montana | 0.063 | 1.824 | 5.412 | 2.882 | 4.412 | 2.529 | 1.235 | 18.987 | 0.323 | 2.53 | 3.62 |
| Wyoming | 0.062 | 0.118 | 1.471 | 0.941 | 1.059 | 2.000 | 0.588 | 6.239 | 0.106 | 0.83 | 1.19 |
| Nevada | 0 | 0 | 0.059 | 0.059 | 0.059 | 0 | 1.706 | 1.883 | 0.032 | 0.25 | 0.36 |
| Utah | 0.040 | 0.059 | 0.765 | 1.176 | 0.647 | 0.176 | 2.706 | 5.569 | 0.095 | 0.74 | 1.06 |
| New Mexico | 0.139 | 0.294 | 0.824 | 0.824 | 0.882 | 0.412 | 2.647 | 6.022 | 0.102 | 0.80 | 1.15 |
| Colorado | 0.115 | 0.294 | 3.765 | 2.059 | 3.294 | 0.824 | 2.176 | 12.527 | 0.213 | 1.67 | 2.39 |
| Arizona | 0.271 | 0.588 | 0 | 0.353 | 0.471 | 0.176 | 3.235 | 5.094 | 0.087 | 0.68 | 0.97 |
| Total: $10^6$ dt/yr | 2.552 | 6.059 | 18.529 | 11.000 | 14.000 | 9.824 | 15.529 | 77.492 | 1.318 | 10.32 | 14.78 |
| Q/yr | 0.043 | 0.103 | 0.315 | 0.187 | 0.238 | 0.167 | 0.264 | — | — | — | — |
| **Total U.S.:** $10^6$ dt/yr | 13.366 | 102.118 | 183.997 | 85.823 | 87.062 | 24.710 | 27.061 | 524.143 | — | 70.63 | 100. |
| Q/yr | 0.228 | 1.736 | 3.128 | 1.459 | 1.480 | 0.420 | 0.460 | 8.911 | — | — | — |

## REFERENCES

1. Saeman, J. F. "Energy and Materials from the Forest Biomass," U.S. Dept. of Agriculture, Forest Service, *Proc. Inst. Gas Tech. Symp. on Clean Fuels from Biomass and Wastes,* USDA, Orlando, FL (1977).
2. Pleeth, S. J. W. *Alcohol, A Fuel for Internal Combustion Engines,* Chapman and Hall, Ltd. (London: 1949).
3. Anderson, L. L. "Energy Potential from Organic Wastes: A Review of the Quantities and Sources," U.S. Bureau of Mines Report No. 8549 (1972).
4. DeRenzo, D. J. "Energy from Bioconversion of Waste Materials," Noyes Data Corp., Park Ridge, NJ (1977).
5. Sharples, J. A. *et al.* "Farm Characteristics, Production, and Land Resources by Production Areas of the North Central Region," Economic Research Service/U.S. Dept. of Agriculture, Statistical Bulletin No. 532 (1974).
6. Clausen, E. C., O. C. Sitton and J. L. Gaddy. "Bioconversion of Crop Materials to Methane," *Proc. Biochem.* (1977).
7. State of California, Dept. of Public Health. *California Solid Waste Planning Study,* Vol. 1 (1968).
8. Inman, R. E. "Silvicultural Biomass Farms," Vol. 1, MITRE Technical Report No. 7347, MITRE Corp./Metrek Div. (1977).
9. Cheremisinoff, P. N. *et al. Woodwastes Utilization and Disposal* (Westport, CT: Technomic Publishing Co., Inc., 1976).
10. American Petroleum Institute, Committee for Air and Water Conservation. "Use of Alcohol in Motor Gasoline—A Review," Publication No. 4082 (1971).
11. Karrer, P. *Organic Chemistry,* 4th ed. (New York: Elsevier North-Holland, 1950). 12.
12. Hagen, D. L. "Methanol—Its Synthesis, Use as a Fuel, Economics, and Hazards," M. S. Thesis, University of Minnesota (1976).
13. Monick, J. A. *Alcohols: Their Chemistry, Properties, and Manufacturing* (Bernhold Book Corp., 1968).
14. Haynes, W. *American Chemical Industry—A History,* Vol. IV (New York: D. Van Nostrand Co., 1948).
15. Woodward, H. F., Jr. "Methanol," R. D. Kirk and N. S. Othmer, Eds. (New York: Wiley-Interscience, 1967). *Encyclopedia of Chemical Technology,* 2nd ed., Vol. 13.

16.  Aharoni, C., and H. Starer. "Adsorption and Desorption of Hydrogen, Carbon Monoxide, and Their Reaction Products on a Catalyst for the Synthesis of Methanol," *Can. J. Chem.* 52 (24) (1974).

17.  Rogerson, P. L. "100-Atm. Methanol Synthesis," *Chem. Eng.* (August, 1973).

18.  Reed, T. B. "Net Efficiencies of Methanol Production from Gas, Coal, Waste, or Wood," in Symposium on *Net Energetics of Integrated Synfuel Systems,* 171st National Meeting, American Chemical Society, Division Fuel Chemistry, New York, Vol. 21, No. 2, April 4-9, 1976.

19.  McGee, R. M. "Co-production of Methanol and SNG from Coal: A Route to Clean Products from Coal Using 'Ready Now' Technology," American Chemical Society, Division Fuel Chemistry, 170th National Meeting, 20(3) (1975).

20.  Reed, T. B. "Biomass Energy Refineries for Production of Fuel and Fertilizer," 8th Cellulose Conf., TAPPI and SUNY, Syracuse, NY, May 20-22, 1975.

21.  Natta, G., P. Pino, G. Mazzanti and I. Pasquon. "Kinetic Interpretations of Heterogeneous Catalysts and Their Applications to Reactions in the Gaseous Phase at High Pressures. I. Synthesis of Methanol," *Chim. Ind. (Milan)* 35 (1953).

22.  Uchida, H., and Y. Ogino. "Rate of Methanol Synthesis," *Bull. Chem. Soc. Japan* 31 (1958).

23.  Vlasenko, V. M., M. G. Rozenfel and M. T. Rusov. "Investigation of Macrokinetics of Synthesis of Methanol and an Industrial Catalyst at High Pressures," *Int. Chem. Eng.* 5 (1965).

24.  Bakemeier, H., P. R. Laurer and W. Schroder. "Development and Application of a Mathematical Model of the Methanol Synthesis," *Methanol Technology and Economics,* G. A. Donner, ed., Chemical Engineering Progress Symposium Series, 66(98) (1970).

25.  Ferraris, G. B., and G. Donati. "Analysis of the Kinetic Models for the Reaction of Synthesis of Methanol," *Quad. Dell Ing. Chim. Ital.* 7(4) (1971).

26.  Kafarov, V. V., V. L. Perov, E. A. Medvedev and V. B. Popov. "Mathematical Modelling and Investigation of the Methanol Synthesis Process over a Low-Temperature Copper-Containing Catalyst," *Int. Chem. Eng.* 15(1) (1975).

27.  Brookhaven National Lab., A.E.C., G. M. Woodwell and E. V. Pecan, Eds. "Carbon and the Biosphere," in *Proc. 24th Brookhaven Symp. in Biol.,* Upton, NY, May 16-18 (1972).

28.  Steinberg, M., M. Beller and J. P. Powell. "A Survey of Applications of Fusion Power Technology to the Chemical and Material Processing Industry," Brookhaven National Laboratory, Department of Applied Science, Report BNL 18866 (1974).

29.  Cappelli, A., A. Collins and M. Dente. "Mathematical Model for Simulating Behavior of Fauser-Montecatini Industrial Reactors for Methanol Synthesis," *Adv. Chem. Series* Vol. 1 (1972).

30. Steinberg, M., and V. Dang. "Use of Controlled Thermonuclear Reactor Fusion Power for the Production of Synthetic Methanol Fuel from Air and Water," Brookhaven National Laboratory, Report BNL 20016 (1975).

31. Williams, K. R., and N. Van Lookeren Campagne. "Synthetic Fuels from Atmospheric Carbon Dioxide," 163rd National Meeting, American Chemical Society, Boston, MA, April 10-14, 1972.

32. Leitz, F. B. "An Electrochemical Carbon Dioxide Reduction Oxygen Generation System Having Only Liquid Waste Products," N. 67-33539 (1967).

33. Stuart, A. K. "Modern Electrolyses Technology," 163rd American Chemical Society National Meeting, Boston, MA, April 9-14, 1972.

34. Gombery, H. J. "Apparatus and Method for Preparing Methanol," *Ger. Offen.* 2(507):407 (1975).

35. Hollander, J. M., and M. Simmons. *Annual Review of Energy,* Vol. I (New York: Annual Reviews, Inc., 1976).

36. Hubbert, M. K. "Energy Resources," in *Resources and Man* (San Francisco: W. H. Freeman & Co., 1969).

37. American Gas Association. "Summary of Gasification Plants," *Coal Age* 80(3) (1975).

38. Bodie, W. W., and K. C. Vyas. "Clean Fuels from Coal," *Oil Gas J.* 72(34) (1974).

39. Thomas, C. O., *et al.* "Methanol from Coal Fuel and Other Applications," Inst. Energy Analysis, Oak Ridge Assoc. Univ., ORAU-126 (IEA75-2) (1976).

40. Mathematical Sciences Northwest, Inc. "Feasibility Study: Conversion of Solid Waste to Methanol or Ammonia," MSNW No. 74-243-1 (1974).

41. Seattle, City of, Dept. of Lighting. "Power Generation Alternatives" (1974).

42. Wiatrek, P. A., *et al.* "Seattle's Solid Waste . . . An Untapped Resource," Seattle Eng. Dept. and Seattle Dept. Lighting (1974).

43. Seattle, City of. "Solid Waste Disposal Incorporating Ferrous Metal Recovery and Production of Methanol or Ammonia," Solid Waste/ Methane/Ammonia Project (1975).

44. Seattle, City of. "Methanol or Ammonia from Municipal Solid Waste" (1975).

45. Seattle, City of. "Methanol or Ammonia from Municipal Solid Waste: A Request for Proposals from Industry," Solid Waste/Methanol/ Ammonia Project (1975).

46. Shah, M. J., and R. E. Stillman. "Computer Control and Optimization of a Large Methanol Plant," *Ind. Eng. Chem.* 62 (1970).

47. Sheehan, R. G. "Methanol from Solid Waste . . . Its Local and National Significance," Eng. Foundation Conf.—Methanol as an Alternate Fuel, Henniker, NH (1974).

48. Sheehan, R. G., and R. Corlett. "Methanol or Ammonia Production from Solid Waste by the City of Seattle," 169th American Chemical Society National Meeting, paper No. 24, April 6-11, 1975.

49. Jordan, R. K., and M. Steinberg. "Applications of Controlled Thermonuclear Reactor (CTR) Fusion Power in the Steel Industry," Brookhaven National Laboratory, BNL 19885 (1975).

50. Hokanson, A. E., and R. M. Rowell. "Methanol from Wood Waste: A Technical and Economic Study," USDA Forest Service, General Technical Report FPL-12 (1977).

51. Anderson, J. E. "Solid Refuse Disposal Process and Apparatus," U.S. Patent No. 3,729,298 (April 24, 1973).

52. Hammond, V. L. "Pyrolysis-Incineration Process for Solid Waste Disposal," Battelle Pacific Northwest Laboratories, Richland, WA (1972).

53. Reed, T. B., and R. M. Lerner. "Methanol: A Versatile Fuel for Immediate Use," *Science,* Vol. 182 (1973).

54. Reed, T. B. "Synthetic Alcohol for Fuel," statement before the Proxmire Committee on Alcohols as Fuel, May 16, 1974.

55. Reed, T. B., R. M. Lerner, E. D. Hingley and R. E. Fahey. "Improved Performance of Internal Combustion Engines Using 5-30% Methanol in Gasoline," MIT, Cambridge, MA (1974).

56. Pefley, R. K., and T. K. Muller. "Letter to Transportation Committee of the California Assembly" (May 21, 1974).

57. Breisacher, P., and R. Nichols. "Fuel Modification: Methanol Instead of Lead as the Octane Booster for Gasoline," Spring Meeting of the Combustion Institute, Madison, WI, March 26-27, 1974.

58. Colucci, J. M. "Methanol Gasoline Blends—Automotive Manufacturer's Viewpoint," General Motors Research Labs., paper presented at the 1974 Engineering Foundation Conference, Henniker, NH, July 7-12, 1974.

59. Ingamells, J. C., and R. H. Lindquist. "Methanol as a Motor Fuel," Chevron Research Co., Richmond, CA (1974).

60. Wigg, E. "Methanol as a Gasoline Extender—Critique," Exxon Research and Eng. Corp. (1975).

61. Baratz, B., R. Quellette, W. Park and B. Stokes. "Survey of Alcohol Fuel Technology," MITRE Corp., Report M74-61 (1975).

62. Starkman, E. S., H. K. Newhall and R. D. Sutton. "Comparative Performance of Alcohol and Hydrocarbon Fuels," Society of Automotive Engineers, Inc., NY (1964).

63. Minomiya, J. S., A. Golovoy and S. S. Labana. "Effect of Methanol on Exhaust Composition of a Fuel Containing Tolune, *n*-Heptane, and Isooctane," *J. Air Poll. Control Assoc.* 20 (1970).

64. Ebersole, G. D. "Power, Fuel Composition, and Exhaust Emission Characteristics of an Internal Combustion Engine Using Isooctane and Methanol," Ph.D. Thesis, University of Tulsa, Tulsa, OK (1971).

65. Adelman, H. G., D. G. Andrews and R. S. Deboto. "Exhaust Emissions from a Methanol-Fueled Automobile," Society of Automotive Engineers —National West Coast Meeting, San Francisco, CA, August 21-24, 1972.

66. Lipinsky, E. S., *et al.* *Fuels from Sugar Crops,* Battelle-Columbus Laboratories, BMI-1957, Columbus, OH (1976).

67. Medville, D., J. Rosenberg, D. Salo and M. Schauffler. "Comparative Economic Assessment of Ethanol from Biomass," MITRE Corp., Report MTR-7936, McLean, VA (1978).

68. Stokes, B., and W. Park. "Survey of Alcohol Fuel Technology," MITRE Corp., Report No. M74-61, McLean, VA (1975).

69. Noyes Development Corp. "Ethyl Alcohol Production Technique," *Symposium of New Developments of Chemical Industries Relating to Ethyl Alcohol, Its By-Products and Wastes,* New York (1964).

70. Clark, D. S., *et al.* "Ethanol from Renewable Resources and Its Application in Automotive Fields," Report of *ad hoc* Alcohol Committee appointed by the Hon. O. E. Lang, Minister of Canadian Wheat Board (1971).

71. Miller, D. L. "Industrial Alcohol from Wheat," paper presented at the Sixth National Conference on Wheat Utilization Research, Oakland, CA, November 5-7, 1969.

72. Miller, D. L. "Fuel Alcohol from Wheat," paper presented at the Seventh National Conference on Wheat Utilization Research, Manhattan, KS, November 3-5, 1971.

73. Miller, D. L. "Agriculture and Industrial Energy," paper presented at the Eighth National Conference on Wheat Utilization Research, Denver, CO, October 10-12, 1973.

74. Scheller, W. A. "Agricultural Alcohol in Automotive Fuel-Nebraska Gasohol," paper presented at the Eighth National Conference on Wheat Utilization Research, Denver, CO, October 10-12, 1973.

75. 16 CFR, Paragraph 1500.14(4).

76. 40 CFR, Paragraph 76.7.

77. Nebraska Code Chapter 66, Paragraph 468.

78. Indiana Bill 1799 (1973).

79. Park, W., G. Price and D. Salo. "Biomass-Based Alcohol Fuels," MITRE Corp., MTR-7866, McLean, VA (1978).

80. Bliss, C., and D. O. Blake. "Silvicultural Biomass Farms: Conversion Processes and Costs," MTR-7347, Vol. V, MITRE/Metrek, McLean, VA (1977).

81. McElroy, A. D. "Utilization of Land with Limited Capabilities," Midwest Research Institute, paper presented at Biomass-A Cash Crop for the Future Conference, Kansas City, MO, March, 1977.

82. Kirk, R. D., and D. S. Othmer, Eds. *Encyclopedia of Chemical Technology,* Vol. 3 (New York: Wiley-Interscience, 1964).

83. Linden, H. R. "A Program for Maximizing Energy Self Sufficiency," *Inst. Gas Technol.* (1974).

84. "Project Independence—An Economic Evaluation," MIT Energy Laboratory, Policy Study Group (1974).
85. "Methanol from Coal can be Competitive with Gasoline," *Oil Gas J.* (December, 1973).
86. "Coal and the Energy Shortage," Continental Oil Co. (1973).
87. "U.S. Energy Outlook—New Energy Forms," National Petroleum Council (1973).
88. Duhl, R. W., and T. O. Wentworth. "Methyl Fuel from Remote Gas Sources," Vulcan-Cincinnati, Inc. (1974).
89. Hearing—U.S. Joint Economic Committee—Subcommittee on Priorities and Economy in the Government, May 20-22, 1974.
90. Jarvis, P. M.   "Methanol as Gas Turbine Fuel," G.E. Gas Turbine Products Division, paper presented at the 1974 Engineering Foundation Conference, Henniker, NH, July 7-12, 1974.
91. McGhee, R. M.   "Methanol Fuel from Natural Gas," paper presented at 1974 Engineering Foundation Conference, Henniker, NH, July 7-12, 1974.
92. Havemann, H. A., *et al.*   "Alcohol in Diesel Engines," *Automobile Eng.* (June 1954).
93. Havemann, H. A., *et al.*   "Alcohol with Normal Diesel Fuels," *Gas Oil Power* Vol. 50 (January 1955).
94. "Firms to Investigate 800-Mile Coal Pipeline," *Mining Eng.* (August 1974).
95. Johnson, J. E.   "The Storage and Transportation of Synthetic Fuels—A Report to the Synthetic Fuels Panel," Oak Ridge National Laboratory, Report ORNL-TM-4307 (1972).
96. "The Handling and Storage of Liquid Propellants," Dept. of the Air Force, Washington, D.C. (1964).
97. Jewell, W. J., *et al.*   "Anaerobic Fermentation of Agricultural Residue: Potential for Improvement and Implementation," New York State College of Agriculture & Life Science, Cornell University, Report EV-76-5-02-2981-7 (1978).
98. Inman, R. E.   "Silvicultural Biomass Farms—Vol. I, Summary," MITRE Tech. Report No. 7347, Metrek Div., McLean, VA (1977).
99. Salo, D. J., R. E. Inman, B. J. McGurk and J. Verhoeff.   "Silvicultural Biomass Farms—Land Suitability and Availability," Vol. III, MITRE Tech. Report No. 7347, Metrek Div., McLean, VA (1977).
100. Odum, E. P.   *Fundamentals of Ecology* (Philadelphia: W. B. Saunders Company, 1959).
101. Geraghty, J. J., D. W. Miller, F. Van Der Leeden and F. L. Troise.   *Water Atlas of the United States,* Water Information Center, Port Washington, NY (1973).
102. *Report of the President's Advisory Panel on Timber and the Environment* (Washington, D.C.: U.S. Government Printing Office, 1973).
103. Dasmann, R.   *Environmental Conservation* (New York: John Wiley & Sons, Inc., 1976).

104. Cotner, M. L. *Land Use Policy and Agriculture: A National Perspecive,* ERS-630, U.S. Dept. of Agriculture, Washington, D.C. (1976).
105. Diderikson, R., A. Hillebaugh and K. Schmude. "Potential Cropland Study," SCS, USDA, Washington, D.C. (1977).
106. Spurr, S. H., and H. J. Vaux. "Timber: Biological and Economic Potential," *Science,* No. 191 (1976).
107. Ledig, F. T., and D. I. H. Linzer. "Fuel Crop Breeding," *Chemtech* (January 1978).

## SELECTED BIBLIOGRAPHY

### GENERAL REFERENCES RELATED TO ALCOHOL
### CHEMISTRY AND PRODUCTION

1. Ahalt, J. D., and J. J. Hairon. "Agriculture in the Seventies—Demand and Price Situation," U.S. Department of Agriculture, Economic Research Service, DPS 124 (1970).
2. Allan, G. G., *et al.* "Pesticides, Pollution and Polymers," *Chem. Technol.* (March 1973).
3. "Chemistry in the Economy," American Chemical Society, Washington, D.C. (1973).
4. Aref, M. M. "Is a Penny a Bushel so Unreasonable," *Agric. Instit. Can. Rev.* 25 (1970).
5. Atomic Energy Commission. "Hydrogen and Other Synthetic Fuels; A Summary of the Work of the Synthetic Fuels Panel," AEC Report TID-26136 (1972).
6. Bare, B. M., and H. W. Lambe. "Economics of Methanol," *Chem. Eng. Prog.* 64(5) (1968).
7. Barry, C. S. "Reduce Claus Sulfur Emissions," *Hydrogen Proc.* 51(4) (1972).
8. Bellar, T. A., and J. E. Sigsby, Jr. "Direct Gas Chromatographic Analysis of Low Molecular Weight Substituted Organic Compounds in Emissions," *Environ. Sci. Techol.* 4 (1970).
9. Blackford, J. L. "Methyl Alcohol (Methanol)," *Chemical Economic Handbook,* Stanford Research Institute (1971).
10. Bohn, H. L. "A Clean New Gas," *Environment* 13 (1971).
11. Bolton, D. H., and D. Hanson. "Economics of Low Pressures in Methanol Plants," *Chem. Eng.* (September 22, 1969).
12. Bryson, E. "Methanol: Old Help for a New Crisis?" *Machine Design* (March 21, 1974).
13. Calvin, M. "Solar Energy by Photosynthesis," *Science* 184:375 (1974); *Proc. 2nd Int. Conf. Single Cell Protein,* MIT, Cambridge, MA (1973).
14. Clewell, H. "Potential of Alcohols as Fuels," statement before the Joint Economic Committee, May 22, 1974.
15. Cobey-Ecco Co. "New High-Volume Fuel Source Available Through Composting Would Ease Energy Crisis," Cobey-Ecco Co., Crestline, OH (February 1974).

115

16.  Cook, L. G. "The Rapidly Changing Technology of Electricity Generation and the Major Consequences in Fossil Fuel Technology," Esso Research and Engineering Co. (1973).
17.  "C79 Methanol Synthesis Catalyst Summary of Laboratory Studies and Commercial Experience," Catalysts and Chemicals, Inc., Louisville, KY (1970).
18.  Danner, G. A. *Methanol Technology and Economics,* G. A. Danner, Ed., Chemical Engineering Progress Symposium Series, 66(98) (1970).
19.  Davis, J. C. "Can Methanol Fuel Contend?" *Chem. Eng.* (June 25, 1973).
20.  Duhl, R. W., and T. O. Wentworth. "Methyl Fuel from Remote Gas Sources," paper presented at Eleventh Annual Meeting, Southern California Section American Institute of Chemical Engineers, Los Angeles (April 16, 1974).
21.  Dutkiewicz, B. "Methanol as a New Energy Source," *Proc. 52nd Annual Conv. of the Nat. Gas Processors Assoc.,* March (1973).
22.  Dutkiewicz, B. "Methanol Competitive with LNG on Long Haul," *Oil Gas J.* (April 30, 1973).
23.  Ganeshan, R. "Methanol as Fuel—Cheaper than LNG," *Oil Gas J.* (July 24, 1972).
24.  Garrett, D., and T. O. Wentworth. "Methyl-Fuel, A New Clean Source of Energy," 1973 Annual Meeting of American Chemical Society, Division of Fuel Chemistry, paper No. 9, August 27, 1973.
25.  Garrett, D. E. "U.S. Synthetic Fuels Development," Garrett Research & Development Co., Western Gas Processors & Oil Refiners Association Annual Fall Meeting, Los Angeles, October 8, 1971.
26.  Harney, B. "Methanol from Coal," U.S. Dept. of the Interior Meeting, February 13, 1974.
27.  Harris, W. D., *et al.* "Methanol from Coal can be Competitive with Gasoline," *Oil Gas J.* (December 17, 1973).
28.  Harvey, H. "Challenges Facing the Petroleum Industry to the Year 2000: An Appraisal," *Fuel* 49 (1970).
29.  Hawkins, P. "Methanol, The Petroleum Situation," Chase Manhattan Bank (1974).
30.  Hedley, B., *et al.* "Methanol: How, Where, Who—Future," *Hydrocarbon Proc.* (September 1970).
31.  Heichel, G. H. "Agricultural Energy Requirements and Food Yield," *Technol. Rev.* (1974).
32.  Heichel, G. H. "Comparative Efficiency of Energy Use in Crop Production," *Conn. Agr. Exp. Sta. Bull.* 739 (1973).
33.  Heichel, G. H. "Energetics of Producing Agricultural Sources of Cellulose," paper presented at the NSF Special Seminar entitled "Cellulose as a Chemical and Energy Resource," University of California, Berkeley, CA, June 26, 1974.
34.  Hill, I. D. "Some Fermentation Forecasts," *Chem. Technol.* (October 1972).

35. Hiller, H., and F. Marschner. "Lurgi Makes Low-Pressure Methanol," *Hydrocarbon Proc.* (September 1970).
36. Hiller, H., *et al.* "The Lurgi Low Pressure Methanol Process," *Chem. Econ. Eng. Rev.* 3 (1971).
37. Hirst, E. "Food-Related Energy Requirements," *Science* 184 (1974).
38. Horsley, J. B., *et al.* "The Design and Performance of the ICI 100 Atmospheres Methanol Plant," paper presented at the AIChE 74th National Meeting and 7th Petroleum & Petrochemicals Exposition, New Orleans, LA, March 11-15, 1973.
39. Hottel, H. C., and J. B. Howard. *New Energy Technology* (Cambridge, MA: MIT Press, 1971).
40. Humphrey, A. E. "Economical Factors in the Assessment of Various Cellulosic Substances as Chemical and Energy Resources," paper presented at the NSF Special Seminar entitled "Cellulose as a Chemical and Energy Resource," University of California, Berkeley, CA, June 26, 1974.
41. Johnson, J. E. "The Storage and Transportation of Synthetic Fuels, A Report to the Synthetic Fuels Panel," ORNL-TM-4307, September (1972).
42. Jones, J. L., and K. S. Vorres. "Clean Fuels from Coal—An Alternative to SNG," The Babcock & Wilcox Co. (1974).
43. Kenard, R. J., and N. M. Nimo. "Present Methanol Manufacturing Costs and Economics Using the ICI Process," Chemical Engineering Progress Symposium Series, 66 (1970).
44. Klass, D. L. "A Perpetual Methane Economy—Is It Possible?" *Chem. Technol.* (March 1974).
45. Manesh, J. S., and R. E. Stillman. "Computer Control and Optimization of a Large Methanol Plant," *Ind. Eng. Chem.* 62 (1970).
46. McLean, J. G. "Coal and the Energy Shortage," Presentation by Continental Oil Co. to Security Analysts, December (1973).
47. "Methanol-for-Fuels on the Horizon," *Chem. Week* (September 19, 1973).
48. Michel, J. W. "Hydrogen and Synthetic Fuels for the Future," paper presented at ACS Symposium entitled "Chemical Aspects of Hydrogen as a Fuel," Chicago, IL, August 27-30, 1973.
49. "Project Independence: An Economic Evaluation," MIT Energy Lab Policy Study Group (1974).
50. Miller, D. L. "Cereal Grains as a Source of Industrial Energy," *Proc. 58th Annual Meeting American Assoc. Cereal Chemists,* St. Louis, MO November 4-8 (1973).
51. Miller, D. L. "Cereals as a Future Source of Industrial Energy," *Proc. Seventh Nat. Conf. on Wheat Utilization Res.,* Manhattan, KS, November 3-5 (1971).
52. Miller, D.L. "Energy From Agriculture," U.S. Energy Outlook, A report by the New Energy Forms Task Group, National Petroleum Council (1973).

53.   Miller, D. L.  "Agriculture and Industrial Energy," paper presented at the 8th Wheat Utilization Research Conference, Denver, CO, October, 1973.

54.   Miller, D. L.  "Ethanol Fermentation and Potential," presented at the NSF Special Seminar entitled "Cellulose as a Chemical and Energy Resource," University of California, Berkeley, June 26, 1974.

55.   Miller, D. L.  "Fuel Alcohol From Wheat," *Proc. Seventh Nat. Conf. on Wheat Utilization Res.,* Manhattan, KS, November (1971).

56.   Mills, G. A., and B. M. Harney.  "Methanol—The 'New Fuel' from Coal," *Chem. Technol.* (January 1974).

57.   National Petroleum Council.  "U.S. Energy Outlook" (December 1972).

58.   "Outlook Bright for Methyl-Fuel," *Environ. Sci. Technol.* (November 1973).

59.   Parker, A.  "World Energy Prospects: An Appraisal," *Fuel* 49 (1970).

60.   Paulsen, T. H.  "Methyl Fuel Project Serves as Attractive Petrochemical Base," *Oil Gas J.* (October 1, 1973).

61.   Perry, R. H., and C. H. Chilton, Eds.  *Chemical Engineers' Handbook,* 5th ed. (New York: McGraw-Hill Book Co., 1973).

62.   Pettman, M. J.  "Redundant Town Gas Plant Now Makes Methanol," *Petrol. Petrochem. Int.* (June 1973).

63.   Pimental, D., *et al.*  "Food Production and the Energy Crisis," *Science* 182 (November 2, 1973).

64.   Pink, J. F.  "Methanol as a Gas Substitute," *Energy Pipelines and Systems* (June 1974).

65.   Quartulli, J.  "Which Route to Bulk Methanol—and at What Cost?" Part I, *Petrol. Petrochem. Int.* (July 1973).

66.   Quartulli, J.  "Which Route to Bulk Methanol—and at What Cost," Part II, *Petrol. Petrochem. Int.* (August 1973).

67.   Quartulli, J.  "Which Route to Bulk Methanol—and at What Cost," Part III, *Petrol. Petrochem. Int.* (September 1973).

68.   Quick, S. L., and G. D. Kittredge.  "Control of Vehicular Air Pollution Through Modification of Conventional Power Plants and Their Fuels," *Proc. Second Int. Clean Air Cong.* (1970).

69.   Ray, D. L.  "The Nation's Energy Future, a Report to R. M. Nixon, President of the United States" (December 1, 1973).

70.   Reed, T. B., and R. M Lerner.  "Methanol: A Versatile Fuel for Immediate Use," *Science* (December 28, 1973).

71.   Reed, T. B., and R. M Lerner.  "Methanol Information Sheet," MIT (January 31, 1974).

72.   Reed, T. B., and R. M Lerner.  "Sources and Methods for Methanol Production," paper presented at The Hydrogen Conference, Miami Beach March 18, 1974.

73.   Reed, T. B.  "A Bibliography on the Production and Use of Alcohols as Fuel," The Methanol Division of the MIT Energy Laboratory (July 1, 1974).

74. Reed, T. B. "Synthetic Alcohol for Fuel," statement before the Joint Economic Committee, May 20, 1974.

75. Rogerson, P. L. "100-Atm. Methanol Synthesis," *Chem. Eng.* (August 20, 1973).

76. Rosenzweig, M., and S. Ushio. "Protein from Methanol," *Chem. Eng.* (January 7, 1974).

77. Royal, M. J. "Why Not Methanol as SNG Feedstock?" *Pipeline Gas J.* (February 1973).

78. Royal, M. J., *et al.* "Big Methanol Plants Offer Cheaper LNG Alternatives," *Oil Gas J.* (February 5, 1973).

79. Saul, J. "Energy for All: We'll Get Tanked Up on Wood Alcohol," Editorial, *Wood & Wood Prod.* (February 1974).

80. Sawicki, E., *et al.* "The 3-Methyl-2-Benzothioazolone Hydrazone —Sensitive New Methods for the Detection, Rapid Estimation, and Determination of Aliphatic Aldehydes," *Anal. Chem.* 33(1) (1961).

81. Scheller, W. A., and B. J. Mohr. "Grain Alcohol-Process, Price and Economic Information," Nebraska Department of Economic Development, Lincoln, NE (June 1974).

82. Schneider, T.R. "Substitute Natural Gas from Organic Materials," ASME Winter Annual Meeting, New York, 1972.

83. Sherwin, M. "Supply/Demand and Process Technology," *Chem. Technol.* (April 1974).

84. Smith, G. "Transco Scans World for Fuel," *New York Times* (September 29, 1973).

85. Soedjanto, P., and F. W. Schaffert. "Transporting Gas—LNG vs. Methanol," *Oil Gas J.* (July 11, 1973).

86. "Statistical Release—Distilled Spirits, June 1973," Bureau of Alcohol, Tobacco, and Firearms, U.S. Department of the Treasury (September 24, 1973).

87. Steinberg, M., *et al.* "Methanol as a Fuel in the Urban Energy Economy and Possible Source of Supply," Brookhaven National Laboratory, Upton, NY (1973).

88. Stephens, G. R., and G. H. Heichel. "Agricultural and Forest Products as Sources of Cellulose," paper presented at the NSF Special Seminar entitled "Cellulose as a Chemical and Energy Resource," University of California, Berkeley, CA, June 26, 1974.

89. Strelzoff, S., "Methanol: Its Technology and Economics," Chem. Eng. Program Symp. Series, 66 (1970).

90. Supp, E. "Technology of Lurgi's Low Pressure Methanol Process," *Chem. Technol.* (July 1973).

91. "Synthetic Organic Chemicals—United States Production and Sales, 1972," United States Tariff Commission, Preliminary (1974).

92. "Synthetic Organic Chemicals—United States Production and Sales, 1971," United States Tariff Commission, TC Publication 614 (1973).

93. Szego, G. C., and C. C. Kemp. "Energy Forests and Fuel Plantations," *Chem. Technol.* (May 1973).

94. Szego, G. A., *et al.* "The Energy Plantation," *Proc. 1972 Intersoc. Energy Conversion Eng. Conf.,* San Diego, CA, September 25-29 (1972).
95. "The Methanol Alternative Now . . . or Never?" *Technol. Rev.* (March/ April 1974).
96. "Two Votes for Methanol," *Chem. Week* (March 14, 1973).
97. "The President's Energy Message," Hearing before the Committee on Interior and Insular Affairs, 92nd Congress, 1st Session, Serial No. 92-1, June 15, 1971.
98. Weikel, T. D. "Ground Support Equipment: Low Pollutant Fuels," Naval Air Engineering Center, Philadelphia, PA (1972).
99. Wentworth, T. "Outlook Bright for Methyl-Fuel," *Environ. Sci. Technol.* 7 (1973).
100. Winter, C. "Energy Imports: LNG vs MeOH," *Chem. Eng.* (November 12, 1973).
101. Yamamoto, T. "Environmental Pollution and Systematization of Chemical Techniques—Use of Methanol as Fuel," *Chem. Econ. Eng. Rev.* 4 (1972).
102. Yamamoto, T. "Synthesis and Utilization of Methanol in Natural Gas Producing Areas (for Food Production)," *Chem. Econ. Eng. Rev.* 5 (1973).

## REFERENCES RELATED TO WASTE UTILIZATION FOR ALCOHOL FUELS AND ENERGY PRODUCTION

1. Abert, J. G., and M. J. Zusman. "Resource Recovery—A New Field for Technology Application," *AiChE J.* 168 (1972).
2. American Public Works Association. *Municipal Refuse Disposal,* 3rd ed., Chicago, IL (1970).
3. Ander, J. E. "The Oxygen Refuse Converter—A System for Producing Fuel, Gas, Oil, Molten Metal and Slag from Refuse," Union Carbide Corp., *ASME Nat. Incinerator Conf. Proc.* (1974).
4. Anderson, J. E. "Union Carbide Oxygen Refuse System," Union Carbide Corp. (1971).
5. Anderson, L. L. "Energy Potential from Organic Wastes: A Review of the Quantities and Sources," U.S. Department of the Interior, Bureau of Mines, I. C. No. 8549 (1972).
6. Appell, H. R., and R. D. Miller. "Fuel from Agricultural Wastes," G.E. Inglett, Symposium: Processing Agricultural and Municipal Wastes, (Westport, CT: The AVI Publishing Co., 1973).
7. Appell, H. R., *et al.* "Conversion of Urban Refuse to Oil," Bureau of Mines, Technical Progress Report No. 25 (May 1970).
8. Appell, H. R., *et al.* "Converting Organic Wastes to Oil: A Replenishable Energy Source," U.S. Department of the Interior, Bureau of Mines, R. I. No. 7560 (1971).

9.   Bailie, R. C., and S. Alpert. "Conversion of Municipal Waste to a Sub-
     stitute Fuel," *Public Works* (August 1973).
10.  "Black Clawson Plant for Hempstead," *Reuse/Recycle* 2(4) (1972).
11.  Boettcher, R. A. "Air Classification of Solid Wastes," EPA Publication
     (SW-30c), report prepared by Stanford Research Institute, Irvine, CA,
     Federal S.W. Program (1972).
12.  Boettner, E. A., *et al.* "Combustion Products from the Incineration of
     Plastic," Environmental Protection Agency, National Technical Infor-
     mation Service Report No. PB-222001 (1973).
13.  Buchbinder, R. I. "A Revolutionary New System for Pollution-Free
     Waste Disposal," Pan American Research, Inc., Annual Meeting Paper
     No. 161 (1973).
14.  Callihan, C. D., and C. E. Dunlap. "Construction of a Chemical-
     Microbial Pilot Plant for Production of SCP from Cellulosic Wastes,"
     SW-24c, U.S. Government Printing Office Stock No. 4402-0027 (1971).
15.  Carver, P. T. "High-Density Compaction Processes for Solid Wastes,"
     *Proc. Third Annual N.E. Regional Antipollution Conf.,* University of
     Rhode Island (Westport, CT: Technomic Publishing Co., Inc., 1970).
16.  Chapman, R. A. "Acid Hydrolysis of Cellulose in Municipal Refuse,"
     Report RC-02-68-11, Public Health Service, Department of Health,
     Education & Welfare (1970).
17.  Cheremisinoff, P. N., *et al.* *Woodwastes Utilization and Disposal*
     (Westport, CT: Technomic Publishing Co., Inc., 1976).
18.  Cheremisinoff, P. N., and A. C. Morresi. *Energy from Solid Wastes*
     (New York: Marcel Dekker, Inc., 1976).
19.  Darney, A., and W. E. Franklin. "Salvage Markets for Materials in
     Solid Wastes," Environmental Protection Agency Report SW-29c
     (1972).
20.  Feldmann, H. F. "Chemical Engineering Applications in Solid Waste
     Treatment," *AIChE Symposium Series 122,* Vol. 68 (1972).
21.  Feldmann, H. F. "Pipeline Gas from Solid Wastes," paper presented at
     the 69th National Meeting, American Institute of Chemical Engineers,
     Cincinnati, OH, 1971.
22.  Feldmann, H. F., *et al.* "Cattle Manure to Pipeline Gas—A Process
     Study," ASME Publication 73-PET-21 (1973).
23.  Glysson, E. A., *et al.* "The Problem of Solid Waste Disposal," Univer-
     sity of Michigan, Ann Arbor, MI (1972).
24.  Golueke, C. G. "Comprehensive Studies of Solid Waste Management,"
     University of California, Third Annual Report, Department of Health,
     Education and Welfare Report SW-105 (1971).
25.  Golueke, C. G. *Composting* (March 1973).
26.  Grethlein, H. E. "The Acid Hydrolysis of Refuse," paper presented at
     the NSF Special Seminar entitled "Cellulose as a Chemical and Energy
     Resource," University of California, Berkeley, CA, June 26, 1974.

27. Halligan, J. E., and R. M Sweazy. "Thermochemical Evaluation of Animal Waste Conversion Processes," *Proc. 72nd Nat. Meeting, Am. Inst. Chem. Eng.,* St. Louis, MO, May (1972).

28. Horner and Shifrin, Inc. "Solid Waste as a Fuel for Power Plants," Environmental Protection Agency Report SW-36d, distributed by the National Technical Information Service, Report No. PB-220-316 (1973).

29. Inglett, G. E. "The Challenge of Waste Utilization," in *Symposium: Processing Agricultural and Municipal Wastes,* G. E. Inglett, Ed. (Westport, CT: The AVI Publishing Co., 1973).

30. Kiange, K-D., *et al.* "Hydrogasification of Cattle Manure to Pipeline Gas," paper presented at the 165th National Meeting, American Chemical Society, Dallas, TX, April 8-13, 1973.

31. Klass, D. L., and S. Ghosh. "Fuel Gas from Organic Wastes," *Chemtech* (November 1973).

32. Klass, D. L., and S. Ghosh. "SNG from Biogasification of Waste Materials," Institute of Gas Technology, paper presented at SNG Symposium I, Chicago, March 12-16, 1973.

33. Liebeskind, J. E. "Pyrolysis for Solid Waste Management," *Chem. Technol.* (September 1973).

34. Loehr, R. C. *Agricultural Waste Management* (New York: Academic Press, Inc., 1974).

35. "Madison Study Focuses on Financial Aspects of Landfilling Milled Trash," *Solid Wastes Managem. Refuse Removal J.* 17(2) (1974).

36. Mallan, G. M. "Preliminary Economic Analysis of the GR&D Pyrolysis Process for Municipal Solid Wastes," Garrett Research and Development Company (1971).

37. "Methanol from Waste," *Newsweek* (January 28, 1974).

38. Miner, J. R. "Farm Animal-Waste Management," Iowa State University of Science and Technology, Agricultural and Home Economics Experiment Stations, Ames, IA, Special Report 67 (1971).

39. Mitchel, D. W. "Growth of Yeast on Refuse Hydrolyzate," M.S. Thesis, Thayer School of Engineering, Hanover, NH (1973).

40. National Center for Resource Recovery. "Materials-Recovery Systems-Engineering Feasibility Study" (1972).

41. Ortuglio, C., *et al.* "Conversion of Municipal and Industrial Refuse Into Useful Materials by Pyrolysis," U.S. Department of the Interior, Bureau of Mines (1973).

42. Pfeffer, J. T., and J. C. Liebman. "Biological Conversion of Organic Refuse to Methane," Semi-Annual Program Report, Department of Civil Engineering, University of Illinois, Report NSF/RANN/SE/61-39191/PR/73/4 (1973).

43. Porteous, A. "The Recovery of Ethyl Alcohol and Protein by Hydrolysis of Domestic Refuse," *Proc. The Institute of Solid Wastes Management Symposium on the Treatment and Recycling of Solid Wastes,* Manchester, England, January 11 (1974).

44.  Reinhardt, J. J., and R. K. Ham. "Final Report on a Demonstration Project at Madison, Wisconsin to Investigate Milling of Solid Wastes," Vol. 1, U.S. Environmental Protection Agency (1973).

45.  *Resource Recovery from Municipal Solid Wastes,* National Center for Resource Recovery, Inc., Washington, D.C. (Lexington, MA:   D. C. Heath & Company, 1974).

46.  Schlesinger, M. D., *et al.* "Pyrolysis of Waste Materials from Urban and Rural Sources," *Proc. Third Mineral Waste Utilization Symp.,* Chicago: IIT Research Institute (1972).

47.  "Seattle's Solid Waste—An Untapped Resource," Department of Engineering and Lighting, City of Seattle (1974).

48.  Sheehan, R. G. "Methanol from Solid Waste . . . Its Local and National Significance," City of Seattle (1974).

49.  Smith, M. L. "Solid Waste Shredding—A Major Change in Waste Control," *Waste Age* (September/October 1973).

50.  "Solid Wastes, An E.S.T. Special Report," *Environ. Sci. Technol.* 4(5) (1970).

51.  "State-of-the-Art Review of Resource Recovery from Municipal Solid Waste," National Center for Resource Recovery (October 1972).

52.  Trezek, G. J., *et al.* "Size Reduction in Solid Waste Processing," 2nd Progress Report—1972-1973, Department of Mechanical Engineering, University of California, Berkeley, CA (1973).

53.  Velzy, C. R., and C. O. Velzy. "The Past, Present and Future of Solid Waste Disposal at Hempstead, New York," *Public Works* (May 1974).

54.  Wilson, D. G. "Energy from Solid Wastes—Needed Government Policy," Testimony before the Subcommittee on Priorities and Economy in Government, Joint Economic Committee, May 20, 1974.

55.  Wilson, D. G.   "Review of Advanced Solid Waste Processing Technology," Paper 40a, presented at the AIChE Symposium on Solid Waste Management, June 4, 1974.

56.  Wilson, D. G., Ed.   *The Treatment and Management of Urban Solid Wastes* (Westport, CT: Technomic Publishing Co., Inc., 1972).

57.  Wise, D. L., *et al.* "Fuel Gas Production from Solid Waste," paper presented at the NSF Special Seminar entitled "Cellulose as a Chemical and Energy Resource," University of California, Berkeley, CA, June 26, 1974.

58.  Yeck, R. G., and P. E. Schleusener. "Recycling of Animal Wastes," *Proc. Nat. Symp. on Animal Waste Managem.,* Graphics Management Corp., Washington, D.C. (1972).

## REFERENCES RELATED TO ALCOHOL-FUELS
## FOR THE AUTOMOBILE

1.  Adelman, H. G., *et al.*   "Exhaust Emissions from a Methanol-Fueled Automobile," National West Coast Meeting of Society of Automotive Engineers, Inc., Paper No. 720693, August 21-24, 1972.

2.  Adt, R. R., Jr., *et al.* "The Hydrogen-Air Fueled Automobile Engine (Part 1)," 8th Intersociety Energy Conversion Engineering Conference, August, 1973.
3.  Adt, R. R., Jr., *et al.* "The Hydrogen and Methanol Air Breathing Automobile Engine," *Proc. Hydrogen Econ. Miami Energy* Conf., University of Miami, March (1974).
4.  Altshuller, A. P. "Vehicular Air Pollution Control," *Gas Scope* (March-April 1970).
5.  Berger, J. E. "Alcohols in Gasoline," statement before the Joint Economic Committee, May 22, 1974.
6.  Billings, R. E., and F. E. Lynch. "History of Hydrogen-Fueled Internal Combustion Engines," and "Performance and Nitric Oxide Control Parameters of the Hydrogen Engine," Energy Research Corp., Provo, Utah, Papers 73001 and 73002 (1973).
7.  Blain, R. "Alexander Graham Bell Recommends Methanol in 1917," *Telephony* (February 25, 1974).
8.  Canada, G. S. "High Pressure Combustion of Liquid Fuels," NASA-CR-134540, NTIS No. N-74-16618 (January 1974).
9.  Carr, R. C., *et al.* "The Influence of Fuel Composition on Emissions of Carbon Monoxide and Oxides of Nitrogen," SAE Paper 700470 (May 1970).
10. Clark, D. S., *et al.* "Ethanol from Renewable Resources and Its Application in Automotive Fuels: A Feasibility Study. Report of a Committee Appointed by the Hon. Otto E. Lang, Minister Responsible for the Canadian Wheat Board," Office of the Minister Responsible for the Canadian Wheat Board, House of Commons, Ottawa, Canada (1971).
11. "Clean Air Car Race, 1970," Clean Air Car Race Organization Committee, Massachusetts Institute of Technology, NTIS No. PB 199479 (1971).
12. Davis, R. "Balancing the Gasohol Triangle," Memo, MIT Lincoln Laboratory (May 1974).
13. DesMarais, A. "Alcohol from Wheat as a Gasoline Additive," Confidential Report of an *Ad Hoc* Committee of the Canadian Science Secretariat, Reported to the Grains Group, Ottawa, Canada (1970).
14. Ebersole, G. "Power, Fuel Consumption, and Exhaust Emission Characteristics of an Internal Combustion Engine using Isooctane and Methanol," Ph.D. Thesis, University of Tulsa, Tulsa, OK (1971).
15. Ebersole, G. D., and F. S. Manning. "Engine Performance and Exhaust Emissions: Methanol versus Isooctane," National West Coast Meeting of Society of Automotive Engineers, Inc., Paper No. 720692, August 21-24, 1972.
16. Eccleston, B. H., *et al.* "Influence of Volatile Fuel Components on Vehicle Emissions," U.S. Dept. of the Interior, Bureau of Mines, Report No. 7291, February (1970).
17. Faeth, G. M., and R. S. Lazar. "Fuel Droplet Burning Rates in a Combustion Gas Environment," *AIAA J.* 9 (1971).

18.  "Feasibility Study of Alternative Fuels for Automotive Transportation
     —Phase I," Esso Research and Engineering Company, Monthly Report
     #2 (August 1972).
19.  Fiala, E. "The Medium and Long Term Adaptation of the Automobile
     Industry to the Changing Energy Situation," paper presented at the
     Hanover Trade Fair, April 24, 1974.
20.  Fitch, F. E., and J. D. Kilgroe. "Investigation of a Substitute Fuel to
     Control Automotive Air Pollution," Consolidated Engineering Tech-
     nology Corp., Contract CPA 22-69-70, NTIS No. PB 194688 (1970).
21.  Fox, T. "Wood Alcohol: Low-Cost Auto Fuel of the Future?"
     *Detroit Free Press* (January 9, 1974).
22.  Glass, W., *et al.* "Evaluation of Exhaust Gas Recirculation for Control
     of Nitrogen Oxide Emissions," Paper 700146, presented at SAE Auto-
     motive Engineering Congress, Detroit, January, 1970.
23.  Gregory, D. P., and R. J. Dufour. "Utilization of Synthetic Fuels
     Other Than Hydrogen," Institute of Gas Technology (1972).
24.  Gregory, D. P., and R. B. Rosenbery. "Synthetic Fuels for Transpor-
     tation and National Energy Needs," paper presented at the Society of
     Automotive Engineers National Meeting Symposium on Energy and the
     Automobile, Detroit, May, 1973.
25.  Handler, P. "Report by the Committee on Motor Vehicle Emissions,"
     National Academy of Sciences (1973).
26.  Heitland, H. "Methanol at Volkswagen," Volkswagen Energy Work-
     shop (1974).
27.  Hirschler, D. A., and F. J. Marsee. "Meeting Future Automobile Emis-
     sion Standards," Paper AM-70-5, presented at National Petroleum
     Refiners Association, San Antonio, TX, April, 1970.
28.  Huls, T. A., and H. A. Nickol. "Influence of Engine Variables on
     Exhaust Oxides of Nitrogen Concentrations from a Multicylinder
     Engine," *Vehicle Emissions—Part III* PT-14, Paper 670482, Society
     of Automotive Engineers (1971).
29.  "Is Alcohol Next Candidate for Fuel Pumps?" *Chem. Week*
     (January 30, 1974).
30.  Jaffee, H., *et al.* "Methanol from Coal for the Automotive Market—
     Draft," AEC Draft Report (1974).
31.  Kant, F. H., *et al.* "Feasibility Study of Alternative Fuels for Auto-
     motive Transportation," Vol. I, II, III, Exxon Research and Engi-
     neering Co. for the Environmental Protection Agency, EPA-46013-
     74-990 (1974).
32.  Mathur, H. B. "Effects of Ethanol Blending on the Performance and
     Exhaust Emission of Spark Ignition Engines," Ph.D. Thesis, Depart-
     ment of Mechanical Engineering, Indian Institute of Technology,
     New Delhi (1971).
33.  McGregor, R., *et al.* "Clean Air Race, 1970, A Summary Report,"
     CACR Committee, MIT (1971).

34.  "Methyl Alcohol as Motor Fuel," Texas Engineering Experimental Station Technical Bulletin, No. 74-2 (April 1974).
35.  "Methyl Fuel Could Provide Motor Fuel," *Chem. Eng. News* (September 17, 1973).
36.  Muzio, L. J., *et al.* "The Effect of Temperature Variations in the Engine Combustion Chamber on Formation and Emission of Nitrogen Oxides," Paper 710158 presented at SAE Automotive Engineering Congress, Detroit, January, 1971.
37.  National Academy of Sciences. "Report by the Committee on Motor Vehicle Emissions," Washington, D.C. (1973).
38.  Newhall, H. K., and I. A. El Messiri. "A Combustion Chamber Design for Minimum Engine Exhaust Emissions," SAE Paper 700491 (1970).
39.  Newhall, H. K., and S. M. Shahed. "Kinetics of Nitric Oxide Formation in High Pressure Flames," *Thirteenth Symposium (International) on Combustion* (1971).
40.  Ninomiya, J. S., *et al.* "Effect of Methanol on Exhaust Composition of a Fuel Containing Toluene, *n*-Heptane and Isooctane," *J. Air Poll. Control Assoc.* 20(5) (1970).
41.  Ogston, A. R. "Alcohol Motor Fuels," *Inst. Petrol. Technol. J.* 23 (1973).
42.  Patterson, D. J., and N. A. Henein. *Emissions from Combustion Engines and Their Control* (Ann Arbor, MI: Ann Arbor Science Publishers, Inc., 1972).
43.  Pefley, R. K., *et al.* "Performance and Emission Characteristics Using Blends of Methanol and Dissociated Methanol as an Automotive Fuel," Paper 719008, Intersociety Energy Conversion Engineering Conference, Boston, MA, August 3-5, 1971.
44.  Pefley, R. K., *et al.* "Study of Decomposed Methanol as a Low Emission Fuel," Final Report to National Air Pollution Control Administration, Contract No. EHS 70-118 (1971).
45.  Pefley, R. K., and T. K. Muller. "Testimony before the California General Assembly Transportation Committee," May 21, 1974.
46.  Reed, T. B., *et al.* "Improved Performance of Internal Combustion Engines Using 5-30% Methanol in Gasoline," Paper for IECEC, San Francisco, August 26-30, 1974.
47.  Sachtschale, J. R. "Exhaust Emission Testing of a 1969 Dodge Operating on Methanol Fuel," Chevron Research Co., Richmond, CA (1970).
48.  Sawhill, J. C. "Potential Use of Ethanol in Automobile Fuels," statement before the Joint Senate Economic Committee, May 21, 1974.
49.  Scheller, W. A. "Agricultural Alcohol in Automotive Fuel-Nebraska Gasohol," paper presented at the 8th National Conference on Wheat Utilization Research, October 10, 1973.
50.  Starkman, E. S., *et al.* "Alternative Fuels for Control of Engine Emissions," *J. Air Poll. Control Assoc.* 20 (1970).

51.  Tillman, R. M.  "Coal-Derived Methanol as Motor Fuel," Continental Oil Co., preprint of paper presented at Henniker, NH Conference entitled "Methanol as an Alternate Fuel," July, 1974.

52.  "Use of Alcohol in Motor Gasoline—A Review," American Petroleum Institute, Publication No. 4082 (1971).

53.  Wenzel, E. C.  "Water/Alcohol Solutions in Internal Combustion Engine Fuel System," Emission Free Fuels, Inc. (1974).

acclimatization—adaptation of a species to a dramatic change in the environment over a period of several generations.

acid gas —gaseous mixture of carbon dioxide ($CO_2$) and hydrogen sulfide ($H_2S$).

activated
sludge —process used to biologically degrade organic matter in dilute water suspension. Diffusion of air through slurry promotes the growth of aerobic bacteria and other organisms, which generates a sludge upon acting on the organic material.

acute toxicity —a poisonous effect caused within a short time period (*e.g.*, 24-96 hr) resulting in severe biological harm or death.

adaptation —change in the structure and/or habit of an organism, resulting in an increase in the overall compatability of the organism with its environment.

adventitious
buds —buds of tissues that develop abnormal positions, *e.g.*, roots growing from stems.

aerobic —refers to a life process that can exist only in the presence of oxygen.

air gasification—a partial combustion process in which coal is burned with roughly 50% of the air needed for complete combustion. The process results in a nonfuel gas, composed of carbon monoxide, a fuel species and nitrogen (has heating value of 1/6 per unit volume of natural gas).

algae —microscopic, simple, marine plant life.

algae bloom —proliferation of algae on the surface of streams, ponds, lakes, etc., population explosion of algae stimulated by phosphate enrichment.

alicyclic —group of organic compounds visualized as an arrangement of carbon atoms in closed-ring structures. The three main subgroups are cycloparaffins (saturated), cycloolefins and cycloacetylenes (cyclynes).

aliphatic —group of organic compounds characterized by straight-chain structures of the constituent carbon atoms. The three main subgroups are paraffins (alkanes), olefins (alkenes or alkadienes) and acetylenes (alkynes). Note that in complex structures their chains can be branched or cross-linked.

alkadeien —olefinic compound, which is comprised of two double bonds that are either adjacent or conjugated.

alkene —aliphatic hydrocarbon compound, consisting of one double bond, *e.g.*, ethylene.

allocthonous
  coal —coal derived from accumulations of plant debris that had been transported from its place of origin and deposited elsewhere.

anaerobic —refers to an organism/life process that cannot exist in the presence of oxygen. Intermediate products of anaerobic digestion are lower-molecular-weight compounds, *e.g.*, alcohols, acids, aldehydes, etc. The end products of anaerobic digestion are methane, carbon dioxide, hydrogen and hydrogen sulfide.

anoxic —refers to an oxygen efficiency.

anthracite —hard coal; hard, black, lustrous coal, comprised of a high percentage of fixed carbon and a low amount of volatiles.

anthraxylon —vitreous-appearing components of coal, derived from the woody tissues of plants, *e.g.*, stems, limbs, twigs, roots. The woody tissues remain as distinct units as coal.

API gravity —American Petroleum Institute standard for measuring the density of oils.

aquatic biota —all living organisms of any designated aquatic area.

aromatic —group of unsaturated cyclic hydrocarbons that consists of one or more rings, *e.g.*, benzene.

autochthonous
  coal —coal originating from accumulations of plant debris located at the place where plants originated.

autotrophs —organisms that use simple chemical species, *e.g.*, iron, sulfur and nitrates, to obtain energy for growth.

auxin —plant hormone or growth substance.

available
  carbon —(free carbon) not chemically bound to oxygen, and therefore, is available for combustion.

banded coal —common variety of bituminous and subbituminous coal composed of a sequence of irregularly alternating layers.

barrel (bbl) —liquid measure of oil (generally crude oil) equivalent to 42 U.S. gallons. One bbl of crude oil is 0.136 tonnes (0.134 long tons and 0.150 short tons).

Bi-Gas —process for coal gasification.

binary cycle —energy recovery system, resulting in heat exchange between two distinct fluid-circulation systems.

bioassay —an assay method that uses a change in biological activity as a means of analyzing a material's response to biological treatment; method of assessing toxic effects of industrial wastes by using viable organisms as test species.

biochemical
 oxygen
 demand —quantity of oxygen used in the biological oxidation of organic matter within a specified time and temperature.
 (BOD)

bioconversion —conversion of one energy form to another via plants or microorganisms, e.g., the synthesis of organic compounds from carbon dioxide by plants, which is the bioconversion of solar energy into stored chemical energy.

biomass —weight of all life in a specified segment of environment; an expression of the total mass of a given population of both plant and animal life.

biosphere —portion of the earth and atmosphere capable of sustaining life.

biosynthesis —production/transformation of substances from other compounds assisted by living organisms.

biota —sum total of living organisms in a designated area.

bitumen —name referring to solid and semisolid hydrocarbons.

bituminous
 coal —soft coal and the most plentiful type. The chief fuel in steam-electrical plants, it is also used as coke for the steel industry and is the raw material of coke by-products, such as various light oils and chemicals.

blended fuel
 oil —mixture of residual and distillate fuel oils.

blending
 naphtha —distillate used as a thinner in processing heavy stocks, e.g., to thin lubricating oils in dewaxing operations.

blending
 stock —stocks used to produce commercial gasoline, e.g., natural gasoline, cracked gasoline, polymer gasoline, aromatics.

British thermal—quantity of heat needed to raise the temperature of 1 lb
 unit (Btu) of water $1°F$ (1 Btu = 1055 J).

butane    —colorless gas composed of 4 carbon and 10 hydrogen atoms used in the manufacture of petrochemicals.

by-products  —secondary products obtained from processing raw material.

calorie    —amount of heat needed to change the temperature of 1 kg of water 1°C.

catalyst    —substance used to speed up a reaction. It does not enter into the actual reaction or the final product.

catalytic
cracking    —(Cat-Craker) process of cracking, in which a catalyst supplements heat and pressure; the conversion of high-boiling hydrocarbons into low-boiling compounds with the use of a catalyst.

chloroplast   —chlorophyll-containing plastid, which serves as the seat of photosynthesis and starch formation.

coal
gasification   —conversion of coal to a gas suitable as a fuel (two commercial processes are Lurgi and Koppers-Totzek).

coal
liquefaction   —(coal hydrogenation) conversion of coal to liquid hydrocarbons and related compounds via hydrogenation.

coal oil    —oil produced from the destructive distillation of bituminous coal.

coal-slurry
pipeline    —pipeline used for the transportation of pulverized coal suspended in water.

coke     —porous, solid residue resulting from the incomplete combustion of coal heated in a closed chamber, *i.e.*, in the absence of air or very little air.

coke-oven gas —gas obtained from coke ovens during the production of coke.

conventional
gas     —natural gas.

crude oil    —unrefined petroleum liquids, straight from the ground.

cyclic
compound   —organic compound whose structure is characterized by a closed ring; its three main groups are alicyclic, aromatic and heterocyclic.

cycloolefin   —(cycloalkene) an alicyclic hydrocarbon with two or more double bonds.

cycloparaffin —an alicyclic hydrocarbon having three or more of its carbon atoms in each molecule united in a ring structure. Each of these ring carbon atoms is bound to two hydrogen atoms or alkyl groups.

| | |
|---|---|
| dehydro-genation | —process in which hydrogen is removed by chemical means. |
| desulfuri-zation | —process in which sulfur and sulfur compounds are removed from gases or liquid hydrocarbon mixtures by chemical or catalytic means. |
| detritus | —dead organic tissues/organisms in an ecosystem. |
| diesel oil | —oil fraction remaining after petroleum and kerosene have been distilled from crude oil. |
| diol | —chemical compound consisting of two hydroxyl groups. |
| ecosystem | —complex of the community of living matter and the environment forming a functioning whole in nature. |
| efficiency, thermal | —fraction or percentage of available energy input that is converted to useful purposes, *i.e.,* Btu output/Btu input. |
| epoxide | —a cyclic ether. |
| ethylene | —colorless, flammable, olefinic gas that has a characteristic sweet color and taste (derived from petroleum cracking). |
| facultative | —having ability to live under severe environmental conditions. |
| firedamp | —mixture of methane and air in coal mines. |
| fixed carbon | —solid residue, other than ash, derived from destructive distillation. |
| food chain | —transfer of nutrients and, hence, energy from one group of organisms to another. |
| fossil fuels | —naturally occurring matter derived from plants and animals; dead organisms that have been transformed into recoverable fuels, including lignite, coal, oil and gas. |
| fuel gas | —synthetic gas used in heating and cooling. |
| fuel oil | —liquefiable petroleum product that can be burned for the generation of heat or power. |
| gas cooling | —process of cooling shifted gas to remove additional hydrocarbon oil by-products and residual phenolic water. |
| gas, natural | —naturally occurring mixture of hydrocarbon gases, found in porous geologic formations beneath earth's crust. The principal constituent of natural gas is methane. |
| gasifications | —refers to the conversion of coal to a high-Btu synthetic natural gas (performed under conditions of high temperatures and pressures). |
| heating value | —heat released by combustion of a unit quantity of a material (measured in calories or Btu). |
| hydrogenation | —process of treating coal with oil and hydrogen under heat and pressure, then separating liquid mixture into products. Typical products are ethane, propane and butane. |

kerosene — petroleum fraction, consisting of hydrocarbons that are slightly heavier than those in gasoline and naphtha.

Krebs cycle — cyclic series of reactions that occur in living organisms, forming a phase of the metabolic function in which acetic acid (or acetyl equivalent) is oxidized through intermediate acids to $CO$ and $H_2O$, thus providing energy for storage in the form of energy-rich phosphate bonds.

liquefaction — process that decomposes and dissolves organic solids via bacterial enzymes outside their cells.

liquified gases — includes ethane, propane, butane, ethylene, propylene and butylenes; obtained from processing natural gas, crude oil and unfinished oils.

liquified natural gas (LNG) — natural gas transformed to a liquid state by cooling to about -260°F, at which point it has a volume of 1/600 of its gaseous volume at normal atmospheric pressure.

methane — lightest among the paraffinic series of hydrocarbons. It is a colorless, odorless gas that is flammable; occurs naturally as swamp gas and is the chief constituent in natural gas.

methanogenesis — process of producing methane.

methogens — bacteria that produce methane.

methylation — replacement of one or more hydrogen atoms with a methyl group.

octane — any of several isomeric liquid paraffin hydrocarbons.

oil shale — sedimentary rock containing solid organic matter from which oil can be extracted upon heating to a high temperature.

olefin (alkene) — group of unsaturated aliphatic hydrocarbons with the general formula $CnH_2n$; contains one or more double bonds and, as such, is chemically reactive. Olefins containing one double bond are called alkenes; those containing two double bonds are called alkadienes.

pentane — any of three isomeric hydrocarbons of the methane group occurring in petroleum.

petrochemicals — chemicals manufactured from components of crude oil and/or natural gas.

phenol — class of aromatic compounds in which one or more hydroxy groups are bound directly to a benzene ring. Examples are phenol, cresols, xylenols, resorcinol and naphthols.

photosynthesis — process in which energy-containing compounds are manufactured from raw materials by green plants when exposed to light.

plankton        —organisms living in the upper portion of any body of water.

propane         —a liquefiable hydrocarbon gas $(C_3H_8)$; a component in raw
                natural gas.

pyrolysis       —transformation of a material into another compound(s) via
                application of heat; also called destructive distillation or the
                heating of organic matter, *e.g.,* coal, wood, petroleum, solid
                wastes, in the absence of oxygen. Initial products are water
                vapor and various volatile compounds. At higher tempera-
                tures, organic matter recombines into complex hydrocarbons
                and water.  Principal products are gases, oils and char.

shift
  conversion    —process in which carbon monoxide in crude gas is converted
                to carbon monoxide to provide the hydrogen:carbon monox-
                ide ratio of 3:1, which is needed for the subsequent synthe-
                sis of methane.

SNG             —synthetic natural gas; manufacture gas consisting roughly of
                97% methane (heating value approximately 1000 Btu/scf).

sour gas        —gas containing hydrogen sulfide.

Syngas          —synthetic gas (SNG).

tail gas        —effluent from a final stage, *e.g.,* as from a flue gas scrubber.

tertiary
  recovery      —the use of heat and methods, exclusive of fluid injection, to
                augment recovery.

translocation   —transfer of metabolites, nutritive material, etc., from one
                part of a plant to another.

underground
  coal gasifi-
  cation        —proposed process for manufacturing synthetic gas from coal
                in natural underground deposits.

vascular plants—plants having specialized conducting systems (xylem and
                phloem).

wet gas         —natural gas containing distillable heavier substances; also
                called casinghead gas; gas containing significant amounts of
                heavy hydrocarbons.

xenobiotics     —matter than can enter the environment only by human
                activities.

zooplankton     —microscopic animal life found floating in lakes, oceans, etc.

# INDEX

137